WAIC and WBIC with R Stan

Joe Suzuki

WAIC and WBIC with R Stan

100 Exercises for Building Logic

 Springer

Joe Suzuki
Graduate School of Engineering Science
Osaka University
Toyonaka, Osaka, Japan

ISBN 978-981-99-3837-7   ISBN 978-981-99-3838-4   (eBook)
https://doi.org/10.1007/978-981-99-3838-4

This Springer imprint is published by the registered company Springer Nature Singapore Pte Ltd.
The registered company address is: 152 Beach Road, #21-01/04 Gateway East, Singapore 189721,
Singapore

Preface: Sumio Watanabe—Spreading the Wonder of Bayesian Theory

The first time I met Prof. Sumio Watanabe was when I was invited to a research meeting organized by Prof. Hideki Aso at the National Institute of Advanced Industrial Science and Technology, where I spoke for about 90 minutes in a seminar. It was in the early summer of 1994, when I had just been appointed as a full-time lecturer at Osaka University, and I believe the topic was about structural learning of Bayesian networks. At that time, there was someone who asked me questions about 20–30 times in total, about once every 2–3 min. That person was Prof. Watanabe.

It was about five years later when Prof. Watanabe gave a lecture titled "Algebraic Geometric Methods in Learning Theory" at the Information Based Induction Sciences (IBIS) Workshop, a machine learning research meeting. At that time, I was also writing papers on algebraic curve cryptography and plane curves (my co-authored paper with J. Silverman has been cited over 100 times), and I was confident in both Bayesian statistics and algebraic geometry. However, I could not understand Prof. Watanabe's IBIS talk at all, as it was too rich in originality.

In conventional Bayesian statistics, regularity (defined in detail in Chap. 2) is assumed, and in that case, the posterior distribution given the sample becomes a normal distribution. Watanabe's Bayesian theory is a generalization of existing Bayesian statistics that uses algebraic geometric methods to derive posterior distributions for cases without assuming regularity. As a consequence, information criteria such as the Widely Applicable Information Criterion (WAIC) and the Widely Applicable Bayesian Information Criterion (WBIC) are derived. These are similar to information criteria such as AIC and BIC, but they can be applied even when the relationship between the true distribution and the statistical model is non-regular. Also, if there is sample data, these values can be easily calculated using software such as Stan (Chap. 3 of this book).

The period from 2005 to 2010 was when Watanabe's Bayesian theory was most developed, and many students entered the Watanabe laboratory. At that time, I listened to several presentations on the achievements of the young researchers in the Watanabe laboratory, but I thought it was impossible to understand without studying the basics. Fortunately, Prof. Watanabe published "Algebraic Geometry and Learning Theory" (Japanese book) in 2006 and "Algebraic Geometry and Statistical Learning Theory"

(Cambridge University Press) in 2009. Both are masterpieces on algebraic geometric methods in learning theory, but they did not discuss the essence of Watanabe's Bayesian theory. The latter book does mention WAIC.

On the other hand, the book "Theory and Methods of Bayesian Statistics" published in 2012 (Japanese book) includes descriptions of not only WAIC but also Watanabe Bayesian theory. However, although it does not assume prior mathematical knowledge, it does not delve into the details of the theory, making it difficult to understand its essence. Honestly, I thought it would be easy to give up unless one spends about a year reading either of the first two books before reading "Theory and Methods of Bayesian Statistics". There was also a concern that the majority of people might be simply believing and using the claim that "use WAIC or WBIC instead of AIC or BIC for non-regular cases". In fact, Watanabe Bayesian theory does not apply to all irregular statistical models. Moreover, with such an attitude, I think it would be difficult to understand the general properties of WAIC and WBIC.

The determination to write this book was solidified when Prof. Watanabe visited the Graduate School of Engineering Sciences at Osaka University for a concentrated lecture in 2019 (he also visited the Osaka University School of Science in 2009). Looking back at the materials from that time, the lecture reminded me more of the general Bayesian theory rather than the essence of Watanabe Bayesian theory. Professor Watanabe avoids difficult topics and shows a caring attitude, but if I were Prof. Watanabe, I would have conveyed the essence of Watanabe Bayesian theory regardless of whether students would run away or not. That thought is also incorporated in this book. At that time, I was planning a series of 100 mathematical exercises on machine learning which was translated into Springer books in English. At that time, without hesitation, I told the editor to include this book as one of them.

However, the book, which I expected to be completed in about half a year, took a full year to finish. I was well aware that Watanabe Bayesian theory is difficult to understand. However, when I started writing, I realized that I only understood the surface of the theory. Moreover, I thought that a satisfactory work would not be completed unless I delved into claims not written in Prof. Watanabe's previous books and papers, and even the essence that Prof. Watanabe himself did not recognize. As I kept questioning why and pursued the issue, I repeatedly encountered new perspectives. I thought that Watanabe Bayesian theory is a masterpiece completed by combining well-thought-out ideas, not just one or two whims.

Watanabe Bayesian theory is constructed by applying algebraic geometry, empirical processes, and zeta functions to existing Bayesian statistics in order to realize generalization without assuming regularity. These are different from linear algebra and calculus and are difficult mathematics that are not used unless you major in mathematics. However, in actual Watanabe Bayesian theory, only a small part of them is used. This book aims to untangle the complex intertwined threads and serve as a guide that allows readers to smoothly understand the material without going through the same time and effort as I did.

The outline of each chapter is as follows. Please see https://bayesnet.org/books for abbreviated solutions to the exercises in each chapter.

Chapter	Contents
2	Overview of Bayesian Statistics
3	The role of MCMC (such as Stan) used in this book and usage of Stan
4	Summary of mathematical matters used in this book
5	Discussions assuming regularity (Watanabe Bayesian theory that corresponds to its generalization)
6	Information criteria (AIC, BIC, TIC, WAIC, WBIC, and Free Energy)
7	Minimal necessary algebraic geometry for understanding Chaps. 8 and 9
8	The essence of WAIC
9	WBIC and its application to machine learning

(The relationship between chapters is illustrated in Fig. 1).
This book has been written with the following features:

1. Covering major topics of Watanabe Bayesian theory, from WAIC/WBIC (Chap. 6) to learning coefficient calculation (Chap. 9). It includes not only the content discussed in Sumio Watanabe's trilogy but also recent results such as the equivalence of WBIC and CV (Cross-Validation) (Chap. 8).
2. Providing R/Stan source code.
3. Presenting numerous examples, making the difficult-to-understand Watanabe Bayesian theory accessible to beginners.
4. Carefully explaining the basics of algebraic geometry necessary for understanding Watanabe Bayesian theory (Chap. 8).
5. Offering 100 exercise problems to allow for self-checks.

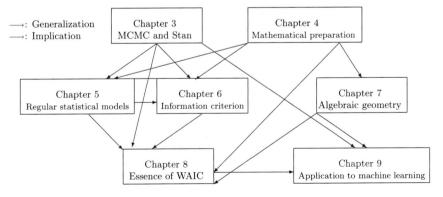

Fig. 1 Relationship between chapters

Additionally, this book is intended for readers who meet any of the following criteria:

1. Possess knowledge equivalent to first-year university-level statistics
2. Have used WAIC and WBIC before (applicable even to non-science majors)
3. Possess knowledge equivalent to "Statistical Learning with Math and R/Python" (Springer).

However, the author would be most delighted if there are readers who have

Studied Watanabe Bayesian theory in the past and failed, but were able to understand it by reading this book

I have tried to avoid making mathematical leaps, but even for previously unknown formulas, the ability to read and understand definitions and explanations is necessary.

My wish is for many people to know about Sumio Watanabe's Bayesian theory, Akaike's information criterion, and Amari's information geometry as some of the great achievements of Japanese statistics.

One-Point Advice for Those Who Struggle with Math

If you struggle with math, get into the habit of "writing". Read this book while writing.

Some people don't take notes when the teacher writes the same thing on the board as the textbook. This reduces the learning effect. On the other hand, writing helps to internalize concepts through sight and touch (sensory and motor nerves). At undergraduate and graduate thesis presentations, math professors often take notes while listening to the student's presentation.

For those who find math difficult while reading this book, I recommend slowly copying the propositions and equations of each chapter. By doing so, the concepts will become your own, and you will want to think about why they are true and consider their proofs. If you still don't understand after copying, try doing the same thing the next day.

This is a study method recommended by the famous mathematician Kunihiko Kodaira. Just by copying, the material becomes much more familiar and less intimidating.

Features of This Series

I have summarized the features of this series rather than this book as follows:

1. **Acquiring: Building Logic**
   By grasping mathematical concepts, constructing programs, executing them, and verifying their operation, readers will build a "logic" in their minds. Not only will you gain knowledge of machine learning but also a perspective that allows you to keep up with new sparse estimation techniques. Most students say that they have learned a lot after solving 100 problems.

2. **Not just talk: Code is available for immediate action**
   It is very inconvenient if there is no source code in a machine learning book. Furthermore, even if there is a package, without the source code, you cannot improve the algorithm. Sometimes, the source is made public on platforms like git, but it may only be available in MATLAB or Python or may not be sufficient. In this book, code is written for most processes, so you can understand what it means even if you don't understand the math.

3. **Not just usage: An academic book written by a university professor**
   Books that only consist of package usage and execution examples have their merits, such as giving people unfamiliar with the topic a chance to grasp it. However, there is a limit to the satisfaction of being able to execute machine learning processes according to procedures without understanding what is happening. In this book, the mathematical principles of each sparse estimation process and the code to implement them are presented, leaving no room for doubt. This book belongs to the academic and rigorous category.

4. **Solving 100 problems: University exercises refined through feedback from students**
   The exercises in this book have been used in university seminars and lectures, refined through feedback from students, and have been carefully selected to be the optimal 100 problems. The main text of each chapter serves as an explanation, so by reading it, you can solve all the exercises.

5. **Self-contained within the book**
   Have you ever been disappointed by a theorem proof that says, "Please refer to literature XX for details"? Unless you are a very interested reader (e.g., a researcher), you are unlikely to investigate the referenced literature. In this book, we have carefully chosen topics to avoid situations where external references are needed. Also, proofs are made accessible, and difficult proofs are placed in the appendix at the end of each chapter.

6. **Not a one-time sale: Videos, online Q&A, and program files**

   In university lectures, we use Slack to answer questions from students 24/365, but in this book, we use the reader's page

   https://bayesnet.org/books

   to facilitate casual interaction between authors and readers. In addition, we publish 10–15 min videos for each chapter. Furthermore, the programs in this book can be downloaded from git.

Toyonaka, Japan   Joe Suzuki

Acknowledgments Finally, we express our profound gratitude to Profs. Sumio Watanabe, Miki Aoyagi, Mr. Yuki Kurumadani, Mr. Ryosuke Shimmura, Mr. Takumi Ikejiri, Mr. Satoshi Kubota, and Ms. Tatsuyoshi Takio of Osaka University for their invaluable cooperation in the creation of this book. We also extend our sincere thanks to Ms. Saki Otani, Mr. Masahiro Suganuma, and Mr. Tetsuya Ishii of Kyoritsu Publishing Co., Ltd. for their assistance, from the planning stage to the editing of this book. A special thank you to Ms. Mio Sugino of Springer for providing the opportunity to publish this manuscript in English.

Contents

Chapter 1
Overview of Watanabe's Bayes

In this chapter, we will first review the basics of Bayesian statistics as a warm-up. In the latter half, assuming that knowledge alone, we will describe the full picture of Watanabe's Bayes Theory. In this chapter, we would like to avoid rigorous discussions and talk in an essay-like manner to grasp the overall picture.

From now on, we will write the sets of non-negative integers, real numbers, and complex numbers as $\mathbb{N}$, $\mathbb{R}$, and $\mathbb{C}$, respectively.

1.1 Frequentist Statistics

For example, let's represent heads of a coin as 1 and tails as 0. If x is a variable representing heads or tails of the coin, x takes the value of 0 or 1. $\mathcal{X} = \{0, 1\}$ is the set of possible values of x. Furthermore, we represent the probability of getting heads with θ that takes values from 0 to 1. $\Theta = [0, 1]$ is the set of possible values of θ. We call θ a parameter. Here, we consider the distribution $p(x|\theta)$ of $x \in \mathcal{X}$ determined by $\theta \in \Theta$. In the case of this coin toss example,

$$p(x|\theta) = \begin{cases} \theta, & x = 1 \\ 1 - \theta, & x = 0 \end{cases} \tag{1.1}$$

can be established. In statistics, when "$p(x|\theta)$ is a distribution", $p(x|\theta)$ must be non-negative and the sum of $x \in \mathcal{X}$ must be 1 (in this case, $p(0|\theta) + p(1|\theta) = 1$).

As it is a coin toss, it might be common to assume that the parameter θ is 0.5. In this case, in statistics, the value of θ is "known". However, if the value of θ is "unknown" because we cannot assume $\theta = 0.5$, for example, due to the coin being bent, we would try tossing the coin several times and estimate the value of θ. If we toss the coin $n = 10$ times and get heads 6 times, we might guess that $\theta = 0.6$, and if we suspect something, we might toss the coin 20 or 100 times.

© The Author(s), under exclusive license to Springer Nature Singapore Pte Ltd. 2023
J. Suzuki, *WAIC and WBIC with R Stan*,
https://doi.org/10.1007/978-981-99-3838-4_1

In this way, the problem of estimating the true value of θ from the data $x_1, \ldots, x_n \in \mathcal{X}$ is called *parameter estimation*, and n (≥ 1) is called the *sample size*. The estimated parameter is often denoted as $\hat{\theta}_n = 0.6$, for example, to distinguish it from the true parameter θ. Alternatively, it can be seen as a mapping like

$$\mathcal{X}^n \ni (x_1, \ldots, x_n) \mapsto \hat{\theta}(x_1, \ldots, x_n) \in \Theta .$$

Statistics is a discipline that deals with problems such as parameter estimation, where the distribution generating $x_1, \ldots, x_n$ is estimated.

In the coin tossing problem, with $x_i = 0, 1$, we have

$$\hat{\theta}(x_1, \ldots, x_n) = \frac{1}{n} \sum_{i=1}^{n} x_i \tag{1.2}$$

which is called the relative frequency. This is the ratio of the frequency of 1 to the number of data n. If $x_1, \ldots, x_n$ occur independently, as shown in Chap. 4, this value converges to the true parameter θ. This is the weak law of large numbers. However, the convergence means that the probability of $|\hat{\theta}(x_1, \ldots, x_n) - \theta|$ staying within a certain value approaches 1, as $x_1, \ldots, x_n$ vary probabilistically.

At this point, it is worth noting that there are two types of averages. The value in (1.2), which is obtained by dividing the sum of the randomly occurring $x_1, \ldots, x_n$ by the number n, is called the *sample mean*. In contrast, $0 \cdot (1 - \theta) + 1 \cdot \theta = \theta$ is called the *expected value*. In this book, when we say average, we mean the latter.

1.2 Bayesian Statistics

However, it is undeniable that estimators like (1.2) can feel awkward. When watching a baseball game, the batting average from the beginning of the season is displayed. By the second half of the season, that average seems to be close to the player's true ability and the true parameter θ. In the opening game, especially in the second at-bat, it is clear that the displayed batting average is either 0 or 1. That is, in the case of $n = 1$, the calculation in (1.2) results in $\hat{\theta}_n = 0$ or $\hat{\theta}_n = 1$. Furthermore, if the first at-bat results in a walk or something that does not count as an at-bat, $n = 0$ and the calculation in (1.2) cannot even be done. Is there a more intuitive estimator? So, considering that there are many hitters around a .250 batting average in baseball, how about estimating as follows?

$$\hat{\theta}(x_1, \ldots, x_n) = \frac{\sum_{i=1}^{n} x_i + 25}{n + 100} . \tag{1.3}$$

The numbers 25 and 100 may be too arbitrary, but they represent prior information and the beliefs of the person making the estimate. The framework that justifies this way of thinking is *Bayesian statistics*. How Eq. (1.3) is derived will be resolved in this chapter.

Before the season begins, there might be someone who imagines the distribution of the player's batting average θ to be roughly around $\theta \in \Theta$. Such a distribution, determined by prior information or the estimator's beliefs, is called the prior distribution, and is denoted by $\varphi(\theta)$. Although it is the same distribution as $p(x|\theta)$, it does not depend on $x \in \mathcal{X}$. However, since it is a distribution, not only must $\varphi(\theta) \geq 0$, but also $\int_\Theta \varphi(\theta)d\theta = 1$. As long as these conditions are met, a uniform distribution such as $\varphi(\theta) = 1, 0 \leq \theta \leq 1$ is acceptable.

The results of the first three at-bats in the opening game, represented by hits as 1 and outs as 0, can be any of the following $(x_1, x_2, x_3) \in \{0, 1\}^3$:

$$000, 001, 010, 011, 100, 101, 110, 111 \; .$$

Here, someone who clearly imagines the prior distribution $\varphi(\theta)$ can calculate the probabilities of these eight events. However, for simplicity, assume that the occurrences of $x_1, x_2, x_3 = 0, 1$ are independent. In fact, the conditional probability for a parameter value of θ is $p(x_1|\theta)p(x_2|\theta)p(x_3|\theta)$. Furthermore, multiplying the prior probability $\varphi(\theta)$ results in $p(x_1|\theta)p(x_2|\theta)p(x_3|\theta)\varphi(\theta)$, but actually, integration over θ is necessary. That is,

$$Z(x_1, x_2, x_3) = \int_\Theta p(x_1|\theta)p(x_2|\theta)p(x_3|\theta)\varphi(\theta)d\theta$$

is the probability of (x_1, x_2, x_3). If the prior distribution is uniform,

$$Z(0, 0, 0) = \int_{[0,1]} (1 - \theta)^3 d\theta = \frac{1}{4} \; , \quad Z(0, 0, 1) = \int_{[0,1]} (1 - \theta)^2 \theta d\theta = \frac{1}{12}$$

$$Z(0, 1, 1) = \int_{[0,1]} (1 - \theta)\theta^2 d\theta = \frac{1}{12} \; , \text{ and } Z(1, 1, 1) = \int_{[0,1]} \theta^3 d\theta = \frac{1}{4}$$

can be calculated in this way. The other four cases can also be calculated similarly, and the sum of the probabilities of the eight sequences is 1. Similarly, for any general $n \geq 1$ and $\varphi(\cdot)$, we can define $Z(x_1, \ldots, x_n)$. This value is called the *marginal likelihood*.

1.3 Asymptotic Normality of the Posterior Distribution

Next, after the opening game is over and a person has seen the results of the first three at-bats (x_1, x_2, x_3), they can estimate the batting average θ more accurately. As the season progresses and a person sees 100 at-bats $(x_1, \ldots, x_{100})$, the estimation

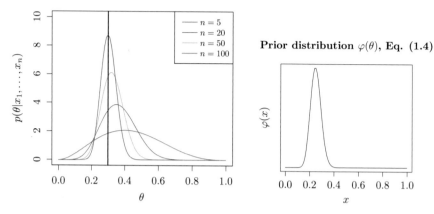

Fig. 1.1 As the sample size increases, the posterior distribution concentrates near the true parameter (left). The prior distribution of batting average (1.4) is maximized at $\theta = 0.25$ (right)

of θ becomes even more accurate. The conditional probability of θ under the data $x_1, \ldots, x_n$ is called its *posterior distribution*. As the sample size n increases, the width of the posterior distribution narrows, concentrating around the true value of θ (Fig. 1.1 left).

To calculate the posterior distribution, it is necessary to apply Bayes' theorem. When expressing the conditional probability of event A under event B as $P(A|B)$, *Bayes' theorem* can be written as

$$P(A|B) = \frac{P(B|A)P(A)}{P(B)}.$$

Let A represent the probability of the parameter being θ, and B represent the probability of the data being $x_1, \ldots, x_n$. That is, by setting $P(B|A)$ as $p(x_1|\theta) \cdots p(x_n|\theta)$, $P(A)$ as the prior probability $\varphi(\theta)$, and $P(B)$ as the marginal likelihood $Z_n(x_1, \ldots, x_n)$, the posterior distribution $p(\theta|x_1, \ldots, x_n)$ can be written as

$$\frac{p(x_1|\theta) \cdots p(x_n|\theta)\varphi(\theta)}{Z(x_1, \ldots, x_n)}.$$

Returning to the batting average example,

$$p(\theta|0, 0, 0) = 4(1 - \theta)^3, \quad p(\theta|0, 0, 1) = 12(1 - \theta)^2\theta,$$

$$p(\theta|0, 1, 1) = 12(1 - \theta)\theta^2, \text{ and } p(\theta|1, 1, 1) = 4\theta^3$$

are the result. The other four cases can also be calculated similarly, and in each of the eight series, $\int_0^1 p(\theta|x_1, x_2, x_3)d\theta = 1$ holds.

As such, defining a prior distribution and finding its posterior probability is Bayesian estimation. In contrast to traditional statistics, which provides a single estimate $\hat{\theta}_n$ for parameter estimation as in (1.2) (called *point estimation*), Bayesian statistics gives the result as a posterior distribution. Under certain conditions (regularity) regarding the relationship between the true distribution and the estimated distribution, it is known that the posterior distribution of the parameter $p(\theta|x_1, \ldots, x_n)$ follows a normal distribution even if it is not a normal distribution when the sample size n is large. This is called *asymptotic normality*.

Regularity $\Longrightarrow$ Asymptotic Normality

This theorem will be proven in Chap. 5, but in practical data analysis using Bayesian statistics, asymptotic normality is often not considered. Rather, it is primarily utilized by some Bayesian theorists to prove mathematical propositions and is often seen as a theory for the sake of theory.

In Watanabe's Bayesian theory, this theorem is generalized. Under the condition of "having a relatively finite variance" defined in Chap. 2, the posterior distribution (asymptotic posterior distribution) is derived when n is large. This posterior distribution generally does not become a normal distribution, but it does become a normal distribution when the regularity condition is added.

On the other hand, if the posterior distribution $p(\theta|x_1, \ldots, x_n)$ is known, the conditional probability $r(x_{n+1}|x_1, \ldots, x_n)$ of $x_{n+1} \in \mathcal{X}$ occurring under the series $x_1, \ldots, x_n \in \mathcal{X}$ can be calculated as

$$r(x_{n+1}|x_1 \ldots, x_n) = \int_\Theta p(x_{n+1}|\theta)p(\theta|x_1, \ldots, x_n)d\theta .$$

This is called the *predictive distribution*. For example, in the case of the batting average problem, where $\mathcal{X} = \{0, 1\}$, the predictive distribution satisfies the properties of a distribution as follows:

$$r(1|x_1, \ldots, x_n) + r(0|x_1, \ldots, x_n) = 1 .$$

That is, when the prior distribution $\varphi(\cdot)$ is determined, the marginal likelihood, posterior distribution, and predictive distribution are also determined. For example, if the prior distribution is

$$\varphi(\theta) = \frac{\theta^{24}(1 - \theta)^{74}}{\int_0^1 \theta_1^{24}(1 - \theta_1)^{74}d\theta_1} , \quad 0 \le \theta \le 1 \tag{1.4}$$

(as shown in Fig. 1.1 on the right), it can be shown that the predictive distribution $r(1|x_1, \ldots, x_n)$ is given by (1.3) (see Sect. 2.1). The *generalization loss*

$$\mathbb{E}_X\left[-\log r(X|x_1, \ldots, x_n)\right] \tag{1.5}$$

and the *empirical loss*

$$\frac{1}{n}\sum_{i=1}^{n}\{-\log r(x_i|x_1, \ldots, x_n)\} \tag{1.6}$$

are also defined using the predictive distribution. These are the mean and arithmetic mean of $-\log r(x|x_1, \ldots, x_n)$ with respect to $x \in \mathcal{X}$, respectively (Chap. 5). The WAIC (Chap. 6) is also defined using the empirical loss.

In other words, Bayesian statistics can be said to be a statistical method that estimates the true distribution using not only samples but also prior distributions that reflect beliefs and prior information.

1.4 Model Selection

In this book, in addition to determining the posterior distribution for parameter estimation, Bayesian statistics are applied for another purpose. Here, we consider the problem of estimating which of the statistical models 0 and 1 is correct from the data sequence $x_1, \ldots, x_n$, where the statistical model (1.1) is model 1 and the statistical model with equal probabilities of 0 and 1 occurring is model 0. In conventional statistics, the details are omitted, but this would typically involve hypothesis testing.

In Bayesian statistics, the value obtained by applying the negative logarithm to the marginal likelihood, $-\log Z(x_1, \ldots, x_n)$, is called the *free energy*. Free energy is used for *model selection*, which estimates which statistical model the sample sequence $x_1, \ldots, x_n$ follows. Under certain conditions, selecting the model with a smaller free energy value results in a correct choice (called *consistency*) as the sample size $n \to \infty$. If the prior probability is uniform at $n = 3$, the marginal likelihood for model 1 can be calculated as

$$-\log Z(0, 0, 0) = \log 4 , \quad -\log Z(0, 0, 1) = \log 12$$

$$-\log Z(0, 1, 1) = \log 12 , \text{ and } -\log Z(1, 1, 1) = \log 4 .$$

In the case of Model 0, regardless of $(x_1, x_2, x_3) \in \mathcal{X}^3$, the free energy becomes $\log 8$, so Model 1 is chosen when $(x_1, x_2, x_3) = (0, 0, 0), (1, 1, 1)$, and Model 0 is chosen otherwise. If we toss a coin three times and the result is biased toward either 0 or 1, rather than a mix of 0 and 1, our intuition tells us that Model 0 is suspicious. The value of the free energy depends on the choice of the prior distribution, but as the sample size n increases, this dependence disappears. In other words, the influence of actual evidence becomes relatively more significant than prior information or the estimator's beliefs.

Next, let's examine how the estimate in (1.2) is obtained. Here, we derive (1.2) from the criterion of maximizing the likelihood. The likelihood is a quantity defined by

$$p(x_1|\theta) \cdots p(x_n|\theta)$$

when data $x_1, \ldots, x_n \in \mathcal{X}$ are obtained. The θ that maximizes this value is called the *maximum likelihood estimator*. The likelihood of (1.2) is $\theta^k(1-\theta)^{n-k}$ when the number of i with $x_i = 1$ is k and the number of i with $x_i = 0$ is $n - k$. We could maximize this value, but instead, we take advantage of the fact that $f(x) = \log x$ is a monotonically increasing function and differentiate

$$k \log \theta + (n - k) \log(1 - \theta)$$

with respect to θ and set it to 0, resulting in

$$\frac{k}{\theta} - \frac{n - k}{1 - \theta} = 0 \,.$$

Solving this equation, we find that (1.2) is obtained.

Furthermore, we assume that the parameter set Θ is a subset of the d-dimensional Euclidean space $\mathbb{R}^d$. Roughly speaking, assuming regularity, it is known that when the sample size n is large, the free energy can be written as

$$\sum_{i=1}^{n} -\log p(x_i|\hat{\theta}_n) + \frac{d}{2} \log n \,,$$

where we have omitted constant terms and written the maximum likelihood estimator obtained from $x_1, \ldots, x_n$ as $\hat{\theta}_n$. This value is called the *BIC*. Also, the dimension d of the parameter space can be interpreted as the number of independent parameters. Furthermore, replacing the second term $\frac{d}{2} \log n$ with d, we obtain

$$\sum_{i=1}^{n} -\log p(x_i|\hat{\theta}_n) + d \,, \tag{1.7}$$

which we call the *AIC*. The details of AIC and BIC will be discussed in Chap. 6. In any case, these criteria are used to select models with smaller values. AIC, BIC, and free energy are examples of quantities used for this purpose, which we call *information criteria*.

1.5   Why are WAIC and WBIC Bayesian Statistics?

As mentioned in the preface, this book is intended for the following readers:

1. Those with a comprehensive knowledge of mathematical statistics.
2. Those who have used WAIC or WBIC but want to understand their essence.
3. Those with a basic understanding of university-level mathematics such as linear algebra, calculus, and probability statistics.

Readers in categories 2 and 3 should be able to approach the level of reader 1 in mathematical statistics by reading up to this point. On the other hand, we often receive questions like the following, particularly from readers in category 2. The purpose of this book is to answer these questions, but we would like to provide an overview of the answers here.

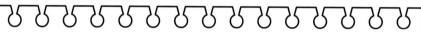

Why are WAIC and WBIC used instead of AIC and BIC in non-regular cases?

Why are WAIC and WBIC considered Bayesian statistics?

Why is algebraic geometry necessary for understanding this?

Information criteria such as AIC, BIC, *WAIC*, and *WBIC* (Chap. 6) can be calculated from the data sequence $x_1, \ldots, x_n$. Here,

$$WAIC = (\text{empirical loss}) + \frac{1}{n} \sum_{i=1}^{n} \mathcal{V}(x_i), \qquad (1.8)$$

where $\mathcal{V}(\cdot)$ is a quantity defined in Chap. 5 and, like empirical loss, is calculated from the posterior distribution $p(\cdot|x_1, \ldots, x_n)$. Calculating this value using R or Python is not difficult.

In any case, by feeding the data sequence into pre-prepared functions, one can calculate the values of such information criteria. However, many people use WAIC instead of AIC in non-regular cases without understanding why, just believing that "in non-regular cases, use WAIC instead of AIC". Just as one would want to understand "why AIC", one would also want to understand "why WAIC". We would like to discuss this in more detail below.

Among these, for WAIC and WBIC, in order to calculate the average of $f : \Theta \to \mathbb{R}$

$$I = \int_{\Theta} f(\theta) p(\theta|x_1, \ldots, x_n) d\theta \qquad (1.9)$$

with respect to the posterior distribution $p(\theta|x_1, \ldots, x_n)$, it is necessary to generate random numbers $\theta = a_1, \ldots, a_m \in \Theta$ ($m \geq 1$) according to $p(\theta|x_1, \ldots, x_n)$, $\theta \in \Theta$

and calculate the approximate value of I:

$$\hat{I} = \frac{1}{m} \sum_{j=1}^{m} f(a_j) \, .$$

Except for special cases, it is considered difficult to calculate the integral of (1.9) mathematically. In this book, we assume the use of specialized software Stan for this purpose. In Chap. 3, we explain how to use it with examples.

Since WAIC and WBIC are calculated by generating random numbers according to the posterior distribution using Stan, it can be inferred that they are Bayesian quantities.

In the following, we denote the operation of taking the average over $x \in \mathcal{X}$ as $\mathbb{E}_X[\cdot]$. For example, the average of $-\log p(x|\hat{\theta}_n)$ over $x \in \mathcal{X}$ is written as $\mathbb{E}_X[-\log p(X|\hat{\theta}_n)]$. Here, we write the variable for which the average is taken in capital letters, such as X. On the other hand, AIC, BIC, and the maximum likelihood estimate $\hat{\theta}_n$ are calculated from the n data points $x_1, \ldots, x_n \in \mathcal{X}$, which actually occur randomly. Therefore, we denote the average of the AIC values taken over these as

$$\mathbb{E}_{X_1 \cdots X_n}[AIC(X_1, \ldots, X_n)] \tag{1.10}$$

or simply as $\mathbb{E}_{X_1 \cdots X_n}[AIC]$. Also, the value of $\mathbb{E}_X[-\log p(X|\hat{\theta}_n)]$ varies depending on the values of $x_1, \ldots, x_n \in \mathcal{X}$ for $\hat{\theta}_n$. Taking the average over these gives

$$\mathbb{E}_{X_1 \cdots X_n} \left[\mathbb{E}_X[-\log p(X|\hat{\theta}(X_1, \ldots, X_n))] \right] , \tag{1.11}$$

which we will simply write this as $\mathbb{E}_{X_1 \cdots X_n} \mathbb{E}_X[-\log p(X|\hat{\theta}_n)]$. Hirotsugu Akaike, who proposed the AIC, considered the value obtained by averaging the maximum log-likelihood $-\log p(x|\hat{\theta}(x_1, \ldots, x_n))$ over both the training data $x_1, \ldots, x_n$ and the test data x (1.11) to be an absolute quantity. He justified the AIC by showing that the AIC averaged over the training data (1.10) matched (1.11).

Sumio Watanabe found it difficult to remove the assumption of regularity as long as the maximum likelihood estimate used in the first term of AIC (1.7) was applied. In Chap. 6, we will prove that the maximum likelihood estimate may not converge to its original value without assuming regularity. In WAIC (1.8), the empirical loss of (1.6) was introduced as a substitute.

> **Why is WAIC Bayesian?**
>
> Breaking away from AIC's maximum likelihood estimation, which does not work in non-regular cases, replacing its first term with the empirical loss was the first step in developing WAIC. As a result, there was a need to explore the derivation of a posterior distribution without assuming regularity.

We have already mentioned that the justification for AIC is that AIC and $\mathbb{E}_X[-\log p(X|\hat{\theta}_n)]$ coincide when averaged over the training data $x_1, \ldots, x_n$. In

Watanabe's Bayesian theory, just as the negative log-likelihood is replaced with the empirical loss, $\mathbb{E}_X[-\log p(X|\hat{\theta}_n)]$ is replaced with the generalization loss of (1.5). Then, with or without regularity, excluding terms that can be ignored when n is large,

$$\mathbb{E}_{X_1\cdots X_n}[WAIC] = \mathbb{E}_{X_1\cdots X_n}[\text{generalization loss}]$$

holds (strictly speaking, the equality does not hold, but the difference becomes negligible). We will discuss the details in Chap. 6 (regular cases) and Chap. 8 (general cases).

Justification of AIC and WAIC ──────────────────────────────

$$\mathbb{E}_{X_1\cdots X_n}[AIC] = \mathbb{E}_{X_1\cdots X_n}\mathbb{E}_X[-\log p(X|\hat{\theta}_n)]$$
$$\mathbb{E}_{X_1\cdots X_n}[WAIC] = \mathbb{E}_{X_1\cdots X_n}[\text{generalization loss}]$$

In regular cases, $WAIC = AIC$

Abandoning maximum likelihood estimation and introducing generalization loss and empirical loss was the starting point of Sumio Watanabe's journey into his novel theory.

1.6 What is "Regularity"

In statistics, assuming the true distribution is $q(x)$ and the statistical model is $p(x|\theta)$, $\theta \in \Theta$, we often use $\mathbb{E}_X[\log \dfrac{q(X)}{p(X|\theta)}]$ to represent the discrepancy between the two. This quantity is called the Kullback-Leibler (KL) information, and it becomes 0 when $p(x|\theta)$ and $q(x)$ match. We will discuss its definition and properties in Chap. 2. Let's denote the value of $\theta \in \Theta$ that minimizes this value as θ_*. Then, the KL information between $p(x|\theta_*)$ and $p(x|\theta)$ is defined as

$$K(\theta) = \mathbb{E}_X[\log \frac{p(X|\theta_*)}{p(X|\theta)}].$$

If this value becomes 0, θ coincides with θ_*, and $p(x|\theta_*)$ is closest to the true distribution $q(x)$. In the following, we will denote the set of such θ_* as Θ_*.

The *regularity* condition (3 conditions), which we have mentioned several times, will be discussed in detail in Chap. 2, but firstly, it requires that Θ_* consists of exactly one element. Secondly, the θ_* contained in Θ_* minimizes $K(\theta)$, but θ_* must not be an endpoint of Θ, and it must not be 0 when differentiated with respect to θ, where Θ is included in the Euclidean space $\mathbb{R}^d$. When we differentiate $K(\theta)$ twice with respect to each of $\theta_1, \ldots, \theta_d$, we denote the matrix with elements $-\dfrac{\partial^2 K(\theta)}{\partial \theta_i \partial \theta_j}$ as

$J(\theta)$. The value of $J(\theta)$ at $\theta = \theta_*$, $J(\theta_*)$, has all non-negative eigenvalues (non-negative definite, details in Chap. 4), but the third condition is that they all take positive values. In regular cases, as n increases, the posterior distribution approaches a normal distribution, and it is shown that the inverse of the covariance matrix is $nJ(\theta_*)$ (Chap. 5). If $J(\theta_*)$ does not have an inverse, the covariance matrix does not exist.

For example, let's assume that the three types of readers mentioned above follow some distribution for each of the three variables: statistics, WAIC/WBIC, and mathematics, and overall, the distribution is the sum of the three distributions divided by 3. The mixture of normal distributions discussed in Chap. 2 corresponds to this case. In this case, it is known that there are multiple θ_* values for which $K(\theta_*) = 0$. Although regular distributions are common in high school and university statistics courses, it is said that most real-world data are non-regular.

1.7 Why is Algebraic Geometry Necessary for Understanding WAIC and WBIC?

In Watanabe's Bayesian theory, the derived posterior distribution is described using the concept of algebraic geometry, specifically the real log canonical threshold λ. It would be impossible to discuss Watanabe's Bayesian theory without using algebraic geometry.

However, Bayesian statistics and algebraic geometry (Chap. 7) are independent academic fields. To solve the problem of generalizing the posterior distribution, a formula called the state density is used to connect the two. The climax could be said to be there.

In Chap. 5, we define $B_n := \{\theta \in \Theta | K(\theta) < n^{-1/4}\}$ and prove that the posterior distribution of $\theta \in \Theta$ not included in B_n can be ignored, regardless of whether it is regular or not. Therefore, the generalization of the posterior distribution to non-regularity in Chap. 8 is performed only for θ included in B_n, i.e., θ close to θ_*.

The state density formula in Chap. 8 seeks an integration formula when $n \to \infty$ and the volume of a certain B_n is sufficiently small (Sect. 8.2). This allows us to express the posterior distribution and free energy using λ and its multiplicity m. The definitions of λ and m are discussed below.

1.8 Hironaka's Desingularization, Nothing to Fear

Many readers may take time to learn the concept of *manifolds* rather than *desingularization*. However, the algebraic geometry covered in Chap. 7 is intended to provide the prerequisite knowledge for Chap. 8 and beyond, and from the perspective of algebraic geometry as a whole, Watanabe's Bayesian theory uses only a very small portion of it.

Hironaka's theorem is a theorem that claims that even if $K(\theta)$ has singular points, there exists a manifold that appears as if there are no singular points. Heisuke Hironaka says the following:

> When you look at the shadow cast by the track of a roller coaster on the ground, an extremely complex figure is drawn there. Lines intersect in various ways, and in some parts, the shape becomes pointed. What was actually a smooth curve appears as intersecting lines and pointed shapes when focusing on the shadow. In algebraic geometry, such points where lines intersect or become pointed are called "singular points".

Let's give two examples of manifolds.

First, it can be easily verified that the set $\mathbb{P}^1$ of the ratio $[x : y]$ of x and y becomes the union of the set U_x of elements that can be written as $[x : 1]$ and the set U_y of elements that can be written as $[1 : y]$. The elements of $U_x \cap U_y$ can be written as both $[x : 1]$ and $[1 : y]$, so we assume a relationship where $xy = 1$. In each case, we have bijections (one-to-one mappings to the top) with $\mathbb{R}^1$ as follows:

$$\begin{cases} \phi_x : U_x \ni [x : 1] \mapsto x \in \mathbb{R}^1 \\ \phi_y : U_y \ni [1 : y] \mapsto y \in \mathbb{R}^1 \end{cases}.$$

Next, notice that the pairs of elements of $\mathbb{P}^1$ and the entire set of real numbers $\mathbb{R}$, as $x, y, z \in \mathbb{R}$, can be written as either $([x : 1], z)$ or $([1 : y], z)$. If we write each set as U_x, U_y, we have bijections with $\mathbb{R}^2$ as follows:

$$\begin{cases} \phi_x : U_x \ni ([x : 1], z) \mapsto (x, z) \in \mathbb{R}^2 \\ \phi_y : U_y \ni ([1 : y], z) \mapsto (y, z) \in \mathbb{R}^2 \end{cases}.$$

If $\mathbb{P}^1$ and $\mathbb{P}^1 \times \mathbb{R}$ satisfy several other conditions, then they are considered to form manifolds of dimensions 1 and 2, respectively. Then, we call the x, y in the first example and the $(x, z), (y, z)$ in the second example the respective *local variables* and their coordinates the *local coordinates*. In both cases, the manifold M is a union of two open sets, but in general, there can be any number of them, and we describe them as a set like $(U_1, \phi_1), (U_2, \phi_2), \ldots$.

From now on, we consider the mapping from $M = \mathbb{P}^1 \times \mathbb{R}$ to $\mathbb{R}^2$

$$g : M \ni ([x : y], z) \mapsto (xz, yz) \in \mathbb{R}^2$$

to be written as $g : (x, z) \mapsto (zx, z)$ when $y \neq 0$, and $g : (y, w) \mapsto (w, wy)$ when $x \neq 0$.

At this stage, let's return to the topic of statistics. Suppose we can express the function $K(\theta)$ for the two-dimensional parameter $\theta = (\theta_x, \theta_y)$ as $\theta_x^2 + \theta_y^2$. When

$\theta_y \neq 0$, we can express (θ_x, θ_y) as (zx, z) and when $\theta_x \neq 0$, we can express (θ_x, θ_y) as (w, wy). In each case, we have

$$K(g(x, z)) = K(zx, z) = z^2(1 + x^2) \,, \quad K(g(y, z)) = K(w, wy) = w^2(1 + y^2) \,. \tag{1.12}$$

Also, for the first case, we obtain

$$\frac{\partial \theta_x}{\partial x} = z \,, \quad \frac{\partial \theta_x}{\partial z} = x \,, \quad \frac{\partial \theta_y}{\partial x} = 0 \,, \quad \frac{\partial \theta_y}{\partial z} = 1$$

and the Jacobian is z for the first case and it is $-w$ for the second case. The absolute values of these are $|z|$ and $|w|$, respectively.

The precise statement of Hironaka's theorem will be discussed in Chap. 7, but it asserts the existence of a manifold M and a mapping $g : M \to \Theta$ such that, for each local coordinate with local variables $u = (u_1, \ldots, u_d)$,

$$K(g(u)) = u_1^{2k_1} \cdots u_d^{2k_d} \tag{1.13}$$

$$|g'(u)| = b(u)|u_1^{h_1} \cdots u_d^{h_d}| \,. \tag{1.14}$$

The form of (1.13) is called *normal crossing*. The absolute value of $g'(u)$ in (1.14) is the Jacobian, and $b(u)$ is a function that always takes positive values. In relation to (1.12), if we replace the local variables (x, z) with (x, v) where $v = z\sqrt{1 + x^2}$, we get

$$K(x, v) = v^2$$

$$|g'(x, v)| = \frac{1}{1 + x^2}|v| \,.$$

In fact, we can calculate

$$\begin{bmatrix} \dfrac{\partial \theta_x}{\partial x} & \dfrac{\partial \theta_x}{\partial v} \\[2mm] \dfrac{\partial \theta_y}{\partial x} & \dfrac{\partial \theta_y}{\partial v} \end{bmatrix} = \begin{bmatrix} \dfrac{1}{(1 + x^2)^{3/2}}u & \dfrac{x}{(1 + x^2)^{1/2}} \\[2mm] -\dfrac{x}{(1 + x^2)^{3/2}}u & \dfrac{1}{(1 + x^2)^{1/2}} \end{bmatrix} \,.$$

On the other hand, if we set $r = w\sqrt{1 + x^2}$, we obtain $K(y, r) = r^2$ and $|g'(y, r)| = \frac{1}{1+y^2}|r|$.

In Watanabe's Bayesian theory, the operation of finding the normal crossing of $K(\theta)$ for each local coordinate is called "resolving the singularity". In other words, for each local coordinate, we obtain two sequences of non-negative integers of length d, $(k_1, \ldots, k_d)$ and $(h_1, \ldots, h_d)$. We define the value

$$\lambda^{(\alpha)} = \min_{1 \le i \le d} \frac{h_i + 1}{2k_i}$$

as the (local-coordinate-wise) real log canonical threshold, and the number of i's achieving the minimum value as the (local-coordinate-wise) multiplicity, $m^{(\alpha)}$.

Then, the minimum of $\lambda^{(\alpha)}$ over all local coordinates, $\lambda = \min_\alpha \lambda^{(\alpha)}$, is called the real log canonical threshold, and the maximum of $m^{(\alpha)}$ among local coordinates achieving $\lambda^{(\alpha)} = \lambda$ is called the multiplicity.

In the example above, for the first local coordinate corresponding to (x, v), the values (k_i, h_i) are $(0, 0)$ and $(2, 1)$, respectively, resulting in $\lambda^{(\alpha)} = 1/2$ and $m^{(\alpha)} = 1$. The same is true for the other local coordinate. Therefore, $\lambda = 1/2$ and $m = 1$.

That is, we can determine λ and m from $K(\theta)$. At this point, the role of algebraic geometry is finished.

Even when using Hironaka's theorem, we are only calculating the normal crossing for each local coordinate. It may not be impossible to understand Watanabe's Bayesian theory by exploring the relationship between statistical regularity and algebraic geometry's singularity, but this book takes the following position:

In Watanabe's Bayesian theory,

the regularity of Bayesian statistics and the non-singularity of algebraic geometry are assumed to be unrelated.

In particular, readers of type 2 should read through Chap. 7 and beyond without the preconception that it is "difficult".

1.9 What is the Meaning of Algebraic Geometry's λ in Bayesian Statistics?

The real log canonical threshold λ in algebraic geometry is also called the *learning coefficient* in Watanabe's Bayesian theory.

Readers of types 1 and 2 who have attempted to study Watanabe's Bayesian theory may have various thoughts about the meaning of λ and how to concretely determine it.

In Chap. 9, we prove that when the system is regular, the normal crossing can actually be calculated, and $\lambda = d/2$. We have mentioned that AIC justifies its validity by satisfying (1.10), but this condition does not hold when the system is not regular. In fact, the expected AIC, $\mathbb{E}_{X_1 \cdots X_n}[AIC]$, becomes smaller. However, even in that case, (1.11) holds. If the system is regular, the second term of WAIC becomes on average equal to d, but when not regular, it becomes 2λ. In fact,

$$\mathbb{E}_{X_1 \cdots X_n}[\text{generalization loss}] = \mathbb{E}_{X_1 \cdots X_n}[\text{empirical loss} + 2\lambda$$

holds (Chap. 8).

However, it is difficult to derive the value of the learning coefficient λ mathematically, and only a few cases have been understood so far (Chap. 9). Among them, Miki Aoyagi's analysis of shrinkage rank regression is famous. In this chapter, we hope that more results will follow, and we have included the proof in the appendix. The proof in the original paper is long, so the author has rewritten it to be simpler and easier to understand the essence.

Instead of the learning coefficient itself, there are research results that seek the upper bound of the learning coefficient. Additionally, there is a method to determine the learning coefficient λ from the WBIC value. The WBIC, using our notation so far, is

$$\sum_{i=1}^{n} - \log p(x_i|\theta)$$

averaged over Θ with respect to the posterior distribution. Watanabe's Bayesian theory generalizes the posterior distribution using an inverse temperature $\beta > 0$. In that case, the WBIC is

$$\int_{\Theta} \sum_{i=1}^{n} - \log p(x_i|\theta) p_\beta(\theta|x_1, \ldots, x_n) d\theta ,$$

where

$$p_\beta(\theta|x_1, \ldots, x_n) = \frac{\varphi(\theta) \prod_{i=1}^{n} p(x_i|\theta)^\beta d\theta}{\int_{\Theta} \varphi(\theta') \prod_{i=1}^{n} p(x_i|\theta')^\beta d\theta'} .$$

WBIC, when regular, exhibits values similar to BIC, and even when not regular, it calculates values close to the free energy. After all, researchers in statistical physics have been attempting to calculate the free energy for a long time, and it is known that the calculations are extensive. WBIC has value as a means to calculate free energy alone.

If one understands the algebraic geometry in Chap. 7 and the state density formula in Sect. 8.1, the WBIC theory is not as complicated as the WAIC theory.

WAIC and WBIC are bifocal glasses

substitutes for AIC and BIC when not regular

Finally, for those who have experienced frustration and want to conquer the challenging Watanabe's Bayesian theory, we have described the important and hard-to-notice essence in blue throughout each chapter. If this helps reach those hard-to-reach spots, we would be delighted.

Chapter 2
Introduction to Watanabe Bayesian Theory

First, we define basic terms in Bayesian[1] statistics, such as prior distribution, posterior distribution, marginal likelihood, and predictive distribution. Next, we define the true distribution q and the statistical model $\{p(\cdot|\theta)\}_{\theta \in \Theta}$, and find the set of $\theta \in \Theta$ that minimizes the Kullback-Leibler (KL) information between them, denoted as Θ_*. We then introduce the concepts of homogeneity with respect to Θ_*, realizability, and regularity between q and $\{p(\cdot|\theta)\}_{\theta \in \Theta}$. The Watanabe Bayes theory aims to generalize the asymptotic normality of the posterior distribution for irregular cases by introducing a condition of relatively finite variance. We derive the relationships between homogeneity, realizability, regularity, and relatively finite variance. Finally, we derive the posterior distribution when the statistical model is an exponential family distribution and its conjugate prior distribution is applied.

2.1 Prior Distribution, Posterior Distribution, and Predictive Distribution

Let $\mathcal{X}$ be a set and $n \geq 1$. Given observed data $x_1, \ldots, x_n \in \mathcal{X}$, we infer information about the unknown distribution q generating them (e.g., the mean and variance). This is called statistical inference. In the following, when we say distribution q, we mean a function q (probability density function) satisfying $q(x) \geq 0$, $x \in \mathcal{X}$, and

$$\int_{\mathcal{X}} q(x)dx = 1 \tag{2.1}$$

---

[1] "Bayes" is derived from Thomas Bayes (1701–1761), a British minister, mathematician, and philosopher.

© The Author(s), under exclusive license to Springer Nature Singapore Pte Ltd. 2023
J. Suzuki, *WAIC and WBIC with R Stan*,
https://doi.org/10.1007/978-981-99-3838-4_2

or if $\mathcal{X}$ is a countable set, a function q satisfying $q(x) \geq 0, x \in \mathcal{X}$, and $\sum_{x \in \mathcal{X}} q(x) = 1$. The integrals treated in this book are Riemann integrals (those learned in high school). The *true distribution* q is the distribution of a *random variable* $X \in \mathcal{X}$ (defined precisely in Chap. 4), and the n samples $x_1, \ldots, x_n$ are the realized values of the independently generated[2] $X_1, \ldots, X_n$. Therefore, the joint distribution of the n samples is the product of each distribution, $q(x_1) \cdots q(x_n)$.

Example 1 For the normal distribution

$$q(x) = \frac{1}{\sqrt{2\pi\sigma^2}} \exp\{-\frac{(x-\mu)^2}{2\sigma^2}\} , \quad \mu \in \mathbb{R} , \ \sigma^2 > 0 , \tag{2.2}$$

we derive (2.1). First, using the substitution integral with $r = (x - \mu)/\sigma$, we note that

$$\int_{-\infty}^{\infty} q(x)dx = \int_{-\infty}^{\infty} \frac{1}{\sqrt{2\pi}} \exp(-\frac{r^2}{2})dr .$$

In general, if $u = r \cos\theta$, $v = r \sin\theta$ ($r > 0, 0 \leq \theta \leq \pi/2$) and $f : \mathbb{R}^2 \to \mathbb{R}$ is differentiable, we can write[3]

$$\int_0^{\infty} \int_0^{\infty} f(u, v)dudv = \int_0^{\pi/2} \int_0^{\infty} f(r \cos\theta, r \sin\theta)r dr d\theta. \tag{2.3}$$

From this and Exercise 1, we obtain

$$\{\int_0^{\infty} \exp(-\frac{u^2}{2})du\}^2 = \frac{\pi}{2} . \tag{2.4}$$

Thus,

$$\int_{-\infty}^{\infty} \exp(-\frac{u^2}{2})du = \sqrt{2\pi} \tag{2.5}$$

is obtained, and (2.1) holds.   ∎

The *mean* and *variance* of a random variable X are defined, respectively, as

---

[2] In this book, uppercase letters like X, X_i represent random variables, and lowercase letters like x, x_i represent their realized values.

[3] When converting the two variables (u, v) to polar coordinates (r, θ), $dudv$ becomes the value obtained by multiplying the Jacobian $\begin{vmatrix} \frac{\partial u}{\partial r} & \frac{\partial v}{\partial r} \\ \frac{\partial u}{\partial \theta} & \frac{\partial v}{\partial \theta} \end{vmatrix} = r$ (polar coordinate transformation).

$$\mathbb{E}[X] := \int_{\mathcal{X}} xq(x)dx \text{ , and}$$

$$\mathbb{V}[X] := \mathbb{E}[\{X - \mathbb{E}[X]\}^2] = \int_{\mathcal{X}} (x - \mathbb{E}[x])^2 q(x)dx \text{ ,}$$

where, when explicitly indicating the random variable X, we will denote it as $\mathbb{E}_X[\cdot]$, $\mathbb{V}_X[\cdot]$. Moreover, for a function $f : \mathcal{X}^n \to \mathbb{R}$, the operation of taking the mean of the random variable $f(X_1, \ldots, X_n)$ is denoted as

$$\mathbb{E}[f(X_1, \ldots, X_n)] := \int_{\mathcal{X}} \cdots \int_{\mathcal{X}} f(x_1, \ldots, x_n)q(x_1) \cdots q(x_n)dx_1 \cdots dx_n.$$

The variance of $f(X_1, \ldots, X_n)$ is defined similarly.

Example 2 For the normal distribution (2.2), the mean and variance of the random variable X are, respectively, $\mathbb{E}[X] = \mu$, $\mathbb{V}[X] = \sigma^2$. In general, $\mathbb{E}[(X - \mu)^m] = 0$ holds for odd numbers m. Furthermore, for $m = 4$, $\mathbb{E}[(X - \mu)^4] = 3\sigma^4$ holds (see Exercise 2). ∎

In general, if the mean of a random variable X is $\mathbb{E}[X]$, the variance of X can be written as

$$\mathbb{V}[X] = \mathbb{E}[\{X - \mathbb{E}[X]\}^2] = \mathbb{E}[X^2] - \mathbb{E}[X\mathbb{E}[X]] - \mathbb{E}[\mathbb{E}[X]X] + \mathbb{E}[X]^2 = \mathbb{E}[X^2] - \mathbb{E}[X]^2.$$

Here, we used $\mathbb{E}[aX] = a\mathbb{E}[X]$ and $\mathbb{V}[bX] = b^2\mathbb{V}[X]$ for $a, b \in \mathbb{R}$. For a d-dimensional vector, similarly, if the mean is $\mathbb{E}[X] \in \mathbb{R}^d$, its covariance matrix can be written as

$$\mathbb{E}[(X - \mathbb{E}[X])(X - \mathbb{E}[X])^\top] = \mathbb{E}[XX^\top] - \mathbb{E}[X]\mathbb{E}[X]^\top \in \mathbb{R}^{d \times d} \text{ .}$$

In the following, the true distribution q is unknown, but we consider finding the distribution closest to the true distribution q among those that can be written as $p(x|\theta)$, $x \in \mathcal{X}$, using some parameter θ. This includes the special case where the true distribution q can be realized in the form of $p(\cdot|\theta)$. We assume that the set of such parameters Θ is contained in a $d(\geq 1)$-dimensional Euclidean space $\mathbb{R}^d$.

Example 3 Suppose that each $x \in \mathcal{X} := \mathbb{R}$ is generated according to the distribution $q(x)$. In this case, if there exists a $\theta = (\mu, \sigma^2) \in \Theta = \{(\mu, \sigma^2) \in \mathbb{R}^2 \mid \sigma^2 > 0\}$ such that $q(x) = p(x|\mu, \sigma^2)$, then the parameters (μ, σ^2) of the normal distribution

$$p(x|\mu, \sigma^2) = \frac{1}{\sqrt{2\pi\sigma^2}} \exp\{-\frac{(x - \mu)^2}{2\sigma^2}\} \tag{2.6}$$

can be estimated from n samples $x_1, \ldots, x_n$. If the true distribution q cannot be written in this way, it will be an approximation. ∎

Furthermore, we assume that each parameter θ is generated according to some distribution φ. That is, we assume that $\varphi(\theta) \geq 0$, $\theta \in \Theta$, and $\int_\Theta \varphi(\theta)d\theta = 1$ hold. In this book, we call $\{p(\cdot|\theta)\}_{\theta \in \Theta}$ a *statistical model* and φ a *prior distribution*.

Then, we call

$$p(\theta|x_1, \ldots, x_n) := \frac{\varphi(\theta)p(x_1|\theta) \cdots p(x_n|\theta)}{Z(x_1, \ldots, x_n)} \tag{2.7}$$

the *posterior distribution* of $\theta \in \Theta$ when $x_1, \ldots, x_n$ are given. The constant $Z(x_1, \ldots, x_n)$ for normalization is called the *marginal likelihood*. We define it as

$$Z(x_1, \ldots, x_n) := \int_\Theta \varphi(\theta)p(x_1|\theta) \cdots p(x_n|\theta)d\theta .$$

We call the *predictive distribution* of $x \in \mathcal{X}$,

$$r(x|x_1, \ldots, x_n) := \int_\Theta p(x|\theta)p(\theta|x_1, \ldots, x_n)d\theta = \frac{Z(x_1, \ldots, x_n, x)}{Z(x_1, \ldots, x_n)} \tag{2.8}$$

which is the average of the statistical model $\{p(\cdot|\theta)\}_{\theta \in \Theta}$ with the posterior distribution (2.7). Note that (2.7) and (2.8) are determined by the prior distribution $\varphi(\cdot)$ and $x_1, \ldots, x_n$. In this book, we will investigate reducing the difference between the true distribution q and the predictive distribution $r(\cdot|x_1, \ldots, x_n)$ as the *sample size* n increases.[4]

The derivation of the marginal likelihood, posterior density, and predictive distribution in Examples 4 to 7 below will be done in Sect. 2.4.

Example 4 In Example 3, the statistical model (2.6) can be written as

$$p(x|\mu) := \frac{1}{\sqrt{2\pi}} \exp\{-\frac{(x-\mu)^2}{2}\}$$

when $\sigma^2 = 1$ is known, and the prior distribution for $\mu \in \Theta = \mathbb{R}$ is

$$\varphi(\mu) := \frac{1}{\sqrt{2\pi}} \exp(-\frac{\mu^2}{2}) . \tag{2.9}$$

In this case, the marginal likelihood and posterior probability are respectively

$$Z(x_1, \ldots, x_n) = \frac{1}{\sqrt{n+1}}(\frac{1}{\sqrt{2\pi}})^n \exp\left\{-\frac{1}{2}\sum_{i=1}^n x_i^2 + \frac{1}{2(n+1)}(\sum_{i=1}^n x_i)^2\right\} \tag{2.10}$$

$$p(\mu|x_1, \ldots, x_n) = \frac{1}{\sqrt{2\pi/(n+1)}} \exp\left\{-\frac{n+1}{2}(\mu - \frac{1}{n+1}\sum_{i=1}^n x_i)^2\right\} . \tag{2.11}$$

---

[4] The number of samples n is called the sample size.

In other words, before and after obtaining the samples $x_1, \ldots, x_n$, the prior distribution $\mu \sim N(0, 1)$ is updated to the posterior distribution $\mu \sim N(\frac{1}{n+1} \sum_{i=1}^{n} x_i, \frac{1}{n+1})$.[5]
Also, the predictive distribution, for $x \in \mathbb{R}$, is

$$r(x|x_1, \ldots, x_n) = \frac{1}{\sqrt{2\pi \dfrac{n+2}{n+1}}} \exp\left\{ -\frac{1}{2\dfrac{n+2}{n+1}} (x - \frac{1}{n+1} \sum_{i=1}^{n} x_i)^2 \right\}. \quad (2.12)$$

That is, it follows a normal distribution with mean $\frac{1}{n+1} \sum_{i=1}^{n} x_i$ and variance $\frac{n+2}{n+1}$. Furthermore, if there exists a μ_* such that $q(\cdot) = p(\cdot|\mu_*)$, when considering $X_1, \ldots, X_n$ as random variables, as $n \to \infty$, $\frac{1}{n+1} \sum_{i=1}^{n} X_i$ converges[6] to μ_*. The derivation of (2.10), (2.11), and (2.12) will be done in Example 17. ∎

Example 5 Consider a statistical model with $\mathcal{X} = 0, 1$, $p(1|\theta) = \theta \in \Theta := [0, 1]$, and suppose there are k occurrences of 1 among $x_1, \ldots, x_n \in \mathcal{X}$. We consider the case where the prior distribution is a uniform distribution:

$$\varphi(\theta) = \begin{cases} 1, & \theta \in \Theta \\ 0, & \theta \notin \Theta \end{cases} \quad (2.13)$$

(we can also confirm that $\varphi(\theta) \geq 0$ and $\int_0^1 \varphi(\theta)d\theta = 1$). In this case, the marginal likelihood, posterior distribution, and predictive distribution are, respectively,

$$Z(x_1, \ldots, x_n) = \frac{(n-k)!k!}{(n+1)!} \quad (2.14)$$

$$p(\theta|x_1, \ldots, x_n) = \theta^k (1-\theta)^{n-k} \frac{(n+1)!}{(n-k)!k!} \quad (2.15)$$

$$r(x|x_1, \ldots, x_n) = \begin{cases} (k+1)/(n+2), & x = 1 \\ (n-k+1)/(n+2), & x = 0 \end{cases}. \quad (2.16)$$

The derivations of (2.14), (2.15), and (2.16) will be done in Example 19. Therefore, if there exists a θ_* such that $q(\cdot) = p(\cdot|\theta_*)$, as $n \to \infty$, the predictive distribution $\frac{k+1}{n+2}$ converges to θ_*. ∎

In Examples 4 and 5, note that specific prior distributions, such as (2.9) and (2.13), were chosen, so the resulting posterior and predictive distributions are obtained accordingly. Different posterior and predictive distributions would be obtained if different prior distributions were chosen.

Example 6 In Example 4, instead of using (2.9) as the prior distribution φ, set

$$\varphi(\mu) := \frac{1}{\sqrt{2\pi}} \exp\{-\frac{(\mu - \phi)^2}{2}\}. \quad (2.17)$$

---

[5] We denote the normal distribution with mean μ and variance σ^2 as $N(\mu, \sigma^2)$.
[6] This is due to the law of large numbers in Sect. 4.3. Here, we mean convergence in probability.

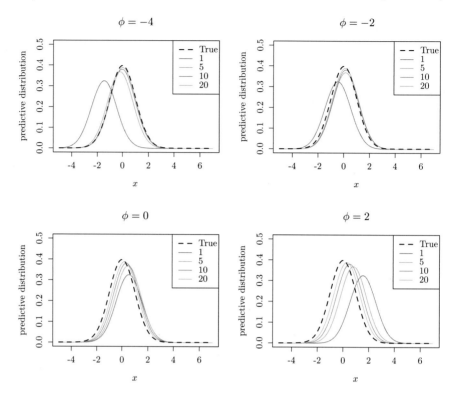

Fig. 2.1 For a specific sequence of random numbers $x_1, \ldots, x_{20}$ generated according to the standard normal distribution, the predictive distribution was constructed using the first $n = 1, 5, 10, 20$ elements. In addition, similar experiments were conducted for $\phi = -4, -2, 2$ as well as $\phi = 0$. It can be seen that the obtained predictive distributions are close to the prior distribution when n is small, but become more dependent on the observed values $x_1, \ldots, x_{20}$ as n increases

Then, the predictive distribution is

$$\frac{1}{\sqrt{2\pi \dfrac{n+2}{n+1}}} \exp\left[-\frac{1}{2\dfrac{n+2}{n+1}}\{x - \frac{1}{n+1}(\phi + \sum_{i=1}^{n} x_i)\}^2\right] \qquad (2.18)$$

(see Fig. 2.1). The derivation of this will be done in Example 18. The case with $\phi = 0$ corresponds to (2.12). The code used for the implementation is shown below. ∎

Please note that the characters in the graphs obtained by executing the programs in this book may differ from those shown in the figure (the code is simplified).

```
1   f <- function(x,mu,sig2) (2*pi*sig2)^(-1/2)*exp(-(x-mu)^2/2)
2   ## Definition of normal distribution
3   phi.seq <- c(-4,2,2) ## Candidate hyperparameters
4   n.seq <- c(1,5,10,20) ## Candidates for n, the first n samples
5   m <- length(phi.seq)
6   l <- length(n.seq)
7   n <- 30 ## Number of samples n
8   x <- rnorm(n) ## Generate n normal random numbers
9   par(mfrow = c(2,2)) ## Generate 4 graphs
10  for(phi in phi.seq){
11      plot(0, 0, xlim=c(-5,7), ylim=c(0,0.5), type="n")
12      for(k in 1:l){
13          nn <- n.seq[k] ## First n samples
14          mu <- (phi+sum(x[1:nn]))/(nn+1) ## Mean of predictive
15      distribution
16          sig2 <- (nn+2)/(nn+1) ## Variance of predictive
17      distribution
18          curve(f(x,mu,sig2), col=k+1, add=TRUE) ## Draw the curve
19          title(paste("phi=",phi))
20      }
21      ## Draw the curve of the true distribution
22      curve(dnorm(x), lwd=2, lty=2, col=1, add=TRUE)
23      legend("topright",c("True", n.seq), lty=c(2, rep(1,4)),
24      lwd=c(2,rep(1,4)), col=1:(1+1))
25  }
26  par(mfrow = c(1,1))
```

Example 7 In Example 5, instead of using (2.13), we use the Beta distribution for φ. That is, with appropriate $a, b > 0$,

$$\varphi(\theta) := \frac{\theta^{a-1}(1-\theta)^{b-1}}{\int_0^1 \theta^{a-1}(1-\theta)^{b-1} d\theta} , \quad \theta \in [0, 1] . \tag{2.19}$$

In this case, the predictive distribution is given by

$$r(x|x_1, \ldots, x_n) = \begin{cases} (k+a)/(n+a+b), & x = 1 \\ (n-k+b)/(n+a+b), & x = 0 \end{cases} . \tag{2.20}$$

The derivation of this result is carried out in Example 19. When $a = b = 1$, this corresponds to (2.16). By making the values of a and b small, the values of (2.20) change sensitively with respect to the observed $k = 0, 1, \ldots, n$. Moreover, if it is expected that there is a high probability of $x = 1$, a prior distribution with $a > b$ should be given.   ∎

Parameters included in the definition of the prior distribution, such as ϕ in (2.17) and a, b in (2.19), are called hyperparameters in this book. The choice of the prior distribution and how to select the hyperparameters cannot be discussed in general without making assumptions.

2.2   True Distribution and Statistical Model

In this book, we consider the problem of estimating the distribution q from observations. However, estimating an unknown q in general is difficult, so we limit ourselves to a certain distribution form $p(\cdot|\theta)$ and estimate q by estimating its $\theta \in \mathbb{R}^d$ $(d \geq 1)$. The true distribution q may or may not be realized using some $\theta \in \Theta$ in $p(\cdot|\theta)$.

$$p(x|\theta) = q(x), \ x \in \mathcal{X}.$$

If there exists a $\theta \in \Theta$ such that the above equation holds, the distribution q is said to be *realizable* by the statistical model $\{p(\cdot|\theta)\}_{\theta \in \Theta}$.

Example 8 Given observed values $x_1, \ldots, x_n \in \mathcal{X}$ generated according to an unknown distribution q, we estimate the mean μ and variance σ^2 using the sample mean $\bar{x} := \frac{1}{n} \sum_{i=1}^{n} x_i$ and the sample variance $s^2 := \frac{1}{n} \sum_{i=1}^{n} (x_i - \bar{x})^2$. Then, using these values, we estimate q with a normal distribution $p(\cdot|\mu, \sigma^2)$, where $\mu := \bar{x}$ and $\sigma^2 := s^2$ (even though q might not be a normal distribution). In this case, $d = 2$ and $\theta = (\mu, \sigma^2)$ is the parameter.   ∎

In general, we want to find the parameter θ that makes $q(\cdot)$ and $p(\cdot|\theta)$ as close as possible. In this book, we evaluate the closeness between the two using the *Kullback-Leibler* (KL) *divergence*

$$D(q||p(\cdot|\theta)) := \int_{\mathcal{X}} q(x) \log \frac{q(x)}{p(x|\theta)} dx .$$

However, the Kullback-Leibler information is not symmetric, as $D(p(\cdot|\theta)||q) \neq D(q||p(\cdot|\theta))$, and thus it is not a distance.

Proposition 1 *When the true distribution is q, for any distribution p, $D(q||p) \geq 0$* and

$$D(q||p) = 0 \iff \text{the probability of } \{x \in \mathcal{X} \mid q(x) = p(x)\} \text{ is } 1$$

holds.

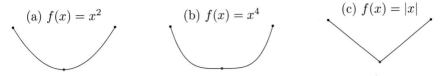

(a) $f(x) = x^2$ (b) $f(x) = x^4$ (c) $f(x) = |x|$

Fig. 2.2 It is not enough for the function to be convex at θ_*. With $\theta_* = 0$, (a) $f(x) = x^2$ satisfies the second condition as $f''(0) = 2$. However, (b) $f(x) = x^4$ is unsuitable due to $f''(0) = 0$. (c) $f(x) = |x|$ is discontinuous with $f'(0-) = -1$ and $f'(0+) = 1$, and $f''(0)$ is not defined

Fig. 2.3 Examples of non-regular cases. From left to right: multiple θ_*, not twice differentiable at θ_*, θ_* on the boundary of Θ

Proof From the inequality $x - 1 \geq \log x$ $(x > 0)$ (equality holds when $x = 1$),

$$\int_x q(x) \log \frac{q(x)}{p(x)} dx = -\int_x q(x) \log \frac{p(x)}{q(x)} dx \geq -\int_x q(x)\{\frac{p(x)}{q(x)} - 1\}dx = 0$$

holds. ∎

In the following, we consider $D(q|p(\cdot|\theta))$ as a function of θ, and the set of θ that minimizes $D(q|p(\cdot|\theta))$ is called the optimal parameter set, denoted as Θ_*. Then, when the following three conditions are satisfied for the statistical model $\{p(\cdot|\theta)\}_{\theta \in \Theta}$ with respect to the distribution q:

1. There exists a (unique) $\theta_* \in \Theta$ such that $\Theta_* = \{\theta_*\}$
2. The matrix[7] $\left[\dfrac{\partial^2 D(q|p(\cdot|\theta))}{\partial \theta_i \partial \theta_j}|_{\theta=\theta_*}\right] \in \mathbb{R}^{d \times d}$ is positive definite
3. There exists an open set $\tilde{\Theta} \subseteq \Theta$ such that $\theta_* \in \tilde{\Theta}$,

the distribution q is said to be *regular* (positive definiteness and open sets are defined in Sect. 4.1).

The second condition requires a certain convexity of the function $D(q||p(\cdot|\theta))$ at $\theta = \theta_*$.[8] For example, if $d = 1$, the condition becomes that the second derivative is positive. Strictly speaking, as shown in Fig. 2.2, it is necessary to confirm that the second derivative is positive or that the Hesse matrix is positive definite, rather than just being convex.

---

[7] A matrix containing the values of second partial derivatives of a multivariate function is called a Hesse matrix (Hesse matrix).

[8] In general, a function $f : \mathbb{R}^d \to \mathbb{R}$ is convex if it satisfies $f(\lambda x + (1 - \lambda)y) \leq \lambda f(x) + (1 - \lambda)f(y)$ for any $0 < \lambda < 1$ and $x, y \in \mathbb{R}^d$.

t-distributions

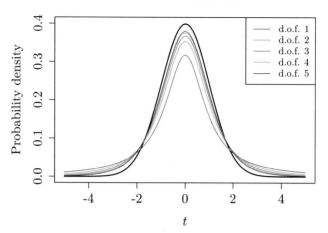

Fig. 2.4 Probability density of t-distributions with degrees of freedom 1, 2, 3, 4, and 5, and the normal distribution (bold black line). As the degrees of freedom of the t-distribution increase, it approaches the normal distribution

The third condition requires that $\theta = \theta_*$, which minimizes $D(q \| p(\cdot | \theta))$, is not on the boundary of Θ. Figure 2.3 schematically represents examples that do not satisfy these three conditions.

Furthermore, we only deal with statistical models that satisfy the following conditions.

Assumption 1 Assume that $p(\cdot | \theta)$ is continuous at $\theta = \theta_*$ in the sense that the following equation holds:

$$\lim_{\theta \to \theta_*} \sup_{x \in \mathcal{X}} | p(x|\theta) - p(x|\theta_*)| = 0 .$$

That is, for any $x \in \mathcal{X}$ and $\theta_* \in \Theta_*$, $p(x|\theta_*)$ is continuous.

In this book, when $p(\cdot | \theta_*) = p(\cdot | \theta_*')$ holds for $\theta_*, \theta_*' \in \Theta_*$, the optimal parameter set Θ_* is said to be *homogeneous*.

Example 9 Let $\Theta = \{\mu \in \mathbb{R} \mid |\mu| \geq 1\}$, and consider

$$q(x) = \frac{1}{\sqrt{2\pi}} e^{-x^2/2}$$

$$p(x|\mu) = \frac{1}{\sqrt{2\pi}} e^{-(x-\mu)^2/2} .$$

In this case,

$$D(q|p(\cdot|\mu)) = \int_{\mathbb{R}} q(x)\{-\frac{x^2}{2} + \frac{(x-\mu)^2}{2}\}dx = -\mu \int_{\mathbb{R}} xq(x)dx + \frac{\mu^2}{2}\int_{\mathbb{R}} q(x)dx = \frac{\mu^2}{2}$$

so, $\Theta_* = \{\pm 1\}$, but Θ_* is not homogeneous. Also, q is neither realizable nor regular with respect to $\{p(\cdot|\mu)\}_{\mu\in\Theta}$. ∎

Example 10 In Example 9, when we set $\Theta = \{\mu \in \mathbb{R} \mid 1 \le \mu \le 2\}$, $\Theta = \{1\}$ is homogeneous. Also, with $d = 1$, the matrix $\left[\frac{\partial^2 D(q|p)}{\partial\theta_i\partial\theta_j}|_{\theta=\theta_*}\right]$ is equal to 1. However, since no open set containing 1 is fully included in Θ, it is not regular. ∎

Example 11 Let $\Theta = \{\mu \in \mathbb{R} \mid |\mu| \le 1\}$, and consider the case where the true distribution is a t-distribution with m degrees of freedom (see Fig. 2.4). In this case, the following calculations can be made:

$$q(x) = \frac{\Gamma(\frac{m+1}{2})}{\sqrt{m\pi}\Gamma(\frac{m}{2})}(1+\frac{x^2}{m})^{-\frac{m+1}{2}}$$

$$p(x|\mu) = \frac{1}{\sqrt{2\pi}}e^{-(x-\mu)^2/2}$$

$$D(q|p(\cdot|\mu)) = \log\left(\frac{\Gamma(\frac{m+1}{2})}{\sqrt{m\pi}\Gamma(\frac{m}{2})}\right) + \frac{1}{2}\log(2\pi) + \mathbb{E}_X[-\frac{m+1}{2}\log(1+\frac{X^2}{m}) + \frac{(X-\mu)^2}{2}],$$

$$\frac{\partial D(q\|p)}{\partial\mu} = \mu, \text{ and}$$

$$\frac{\partial^2 D(q\|p)}{\partial\mu^2} = 1 > 0.$$

Although the true distribution q is not realizable in the statistical model $\{p(\cdot|\mu)\}_{\mu\in\Theta}$, $D(q|p(\cdot|\mu))$ is minimized only when $\mu = 0$, and there exists an $\epsilon > 0$ such that $0 \in (-\epsilon, \epsilon) \subseteq \Theta$, so it is regular. ∎

Example 12 Let $\Theta = \{(\alpha, \beta) \in \mathbb{R}^2 | \alpha \neq 0\}$, $\mu \in \mathbb{R}$, $\sigma > 0$, and consider

$$q(x) = \frac{1}{\sqrt{2\pi\sigma^2}}e^{-(x-\mu)^2/2\sigma^2},$$

$$p(x|\alpha, \beta) = \frac{1}{\sqrt{2\pi/\alpha^2}}e^{-(\alpha x - \beta)^2/2}.$$

In this case, only when $(\alpha, \beta) = (\pm\frac{1}{\sigma}, \pm\frac{\mu}{\sigma})$ (with the same order of signs),

$$D(q|p) = \int_{\mathbb{R}} q(x) -\frac{1}{2}\log\sigma^2\alpha^2 - \frac{(x-\mu)^2}{2\sigma^2} + \frac{(\alpha x - \beta)^2}{2}dx = 0$$

is achieved. Also, since $p(x|\frac{1}{\sigma}, \frac{\mu}{\sigma}) = p(x| -\frac{1}{\sigma}, -\frac{\mu}{\sigma})$, $\Theta_* = \{(\frac{1}{\sigma}, \frac{\mu}{\sigma}), (-\frac{1}{\sigma}, -\frac{\mu}{\sigma})\}$ is homogeneous. ∎

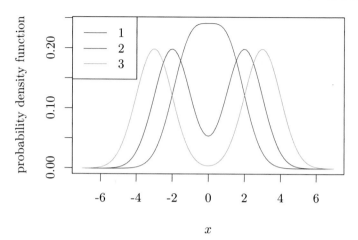

Fig. 2.5 Mixture of normal distributions with parameters $(\mu, -\mu, 0.5)$, $\mu = 1, 2, 3$

Example 13 In Example 12, if we set $\Theta = \{(\alpha, \beta) \in \mathbb{R}^2 \mid \alpha > 0\}$, $\mu = 0$, and $\sigma^2 = 1$, then only when $(\alpha, \beta) = (1, 0)$, $D(q\|p) = 0$ can be achieved. Then,

$$\frac{\partial^2 D(q\|p)}{\partial \alpha^2} = \frac{\partial^2}{\partial \alpha^2} \int_{\mathbb{R}} \{-\log \alpha + \frac{(\alpha x - \beta)^2}{2}\} q(x) dx = \int_{\mathbb{R}} \{\frac{1}{\alpha^2} + x^2\} q(x) dx = 2 \ .$$

$$\frac{\partial^2 D(q\|p)}{\partial \alpha \partial \beta} = \frac{\partial^2}{\partial \alpha \partial \beta} \int_{\mathbb{R}} \frac{(\alpha x - \beta)^2}{2} q(x) dx = 0 \ , \text{ and}$$

$$\frac{\partial^2 D(q|p)}{\partial \beta^2} = \frac{\partial^2}{\partial \beta^2} \int_{\mathbb{R}} \frac{(\alpha x - \beta)^2}{2} q(x) dx = 1$$

hold, and the matrix $\left[\frac{\partial^2 D(q|p)}{\partial \theta_i \partial \theta_j} |_{\theta = \theta *} \right] = \begin{bmatrix} 2 & 0 \\ 0 & 1 \end{bmatrix}$ has both eigenvalues positive. Furthermore, if we take a circle with radius $\epsilon > 0$ centered at $(1, 0)$ small enough, that circle is included in Θ. Therefore, q is regular with respect to $\{p(\cdot|\alpha, \beta)\}_{(\alpha, \beta) \in \Theta}$. ∎

Let's give a slightly more specific example.

Example 14 *(Mixture of normal distributions* [12]*)* Consider a distribution with the probability density function given by

$$p(x|a, \mu_1, \mu_2) = (1 - a) \cdot \frac{1}{\sqrt{2\pi}} \exp(-\frac{(x - \mu_1)^2}{2}) + a \cdot \frac{1}{\sqrt{2\pi}} \exp(-\frac{(x - \mu_2)^2}{2}).$$

Here, $0 < a < 1$, and it is a mixture distribution with the ratio of $1 - a, a$ for the probability density functions of normal distributions $N(\mu_1, 1)$ and $N(\mu_2, 1)$ (the graph for the case of $a = 0.5$, $\mu_1 = \mu$, $\mu_2 = -\mu$, $\mu = 1, 2, 3$ is shown in Fig. 2.5). In the following, we consider the case where $\mu_1 = 0$, $\mu_2 = b$, and $\Theta =$

$\{\theta = (a, b) \mid 0 \le a \le 1, -\infty \le b \le \infty\}$, $q(x) = p(x|a_*, 0, b_*)$. In this case, the Kullback-Leibler divergence $D(q||p)$ is

$$K(a, b) := \mathbb{E}[\log \frac{(1 - a_*) \exp(-x^2/2) + a_* \exp(-(x - b_*)^2/2)}{(1 - a) \exp(-x^2/2) + a \exp(-(x - b)^2/2)}]$$

$$= \int_{\mathcal{X}} \log \left(\frac{1 + a_*\{\exp(b_* x - b_*^2/2) - 1\}}{1 + a\{\exp(bx - b^2/2) - 1\}} \right) q(x)dx .$$

However, when $a_* b_* = 0$, $ab = 0 \Longleftrightarrow K(a, b) = 0$. That is, $\theta = (a, b) \in \Theta$ satisfying $K(a, b) = 0$ is not unique. ∎

2.3 Toward a Generalization Without Assuming Regularity

Next, for $p(\cdot|\theta_*)$, $\theta_* \in \Theta_*$, and $p(\cdot|\theta)$, $\theta \in \Theta$, we define the log-likelihood ratio as

$$\log \frac{p(\cdot|\theta_*)}{p(\cdot|\theta)} . \tag{2.21}$$

Then, if there exists a constant $c > 0$ such that for any $\theta_* \in \Theta_*$ and $\theta \in \Theta$,

$$\mathbb{E}_X[\{\log \frac{p(X|\theta_*)}{p(X|\theta)}\}^2] \le c\mathbb{E}_X[\log \frac{p(X|\theta_*)}{p(X|\theta)}] \tag{2.22}$$

holds, then for the statistical model $\{p(\cdot|\theta)\}_{\theta \in \Theta}$ and the true distribution q, the log-likelihood ratio function is said to have *relatively finite variance*.

In Chap. 5, we show that when regularity can be assumed, the posterior probability converges in law to a normal distribution (the definition is given in Chap. 4). Watanabe's Bayesian theory provides a solution for the law of convergence of the posterior distribution in a general situation without assuming regularity. In Chap. 8, we will use the law of convergence of a certain empirical process to a Gauss process (a generalization of the central limit theorem) to derive the conclusion, but the condition of relatively finite variance is essential for enabling its application. WAIC, WBIC (second half of Chap. 8), and the calculation of learning coefficients (Chap. 9) are obtained as applications.

Proposition 2 (Watanabe 2012 [13])

1. *Having relatively finite variance* $\Longrightarrow \Theta_*$ *is homogeneous*
2. *If q is realizable with respect to* $\{p(\cdot|\theta)\}_{\theta \in \Theta}$, $\Longrightarrow$ *it has relatively finite variance*
3. *If q is regular with respect to* $\{p(\cdot|\theta)\}_{\theta \in \Theta}$, $\Longrightarrow$ *it has relatively finite variance*

Proof The knowledge of matrix and mean value theorems in Chap. 4 is required, so the proof is provided in the appendix of Chap. 4.

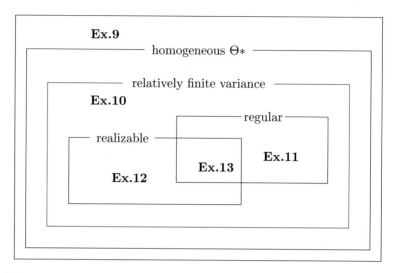

Fig. 2.6 The relationship between $\{p(\cdot|\theta)\}_{\theta\in\Theta}$, regularity, realizability, relatively finite variance, and homogeneous $\Theta*$ in Examples 9–13

Example 15 In the following, we illustrate Proposition 2 through Examples 9–13 (Fig. 2.6). First, for Examples 9 and 10, we can transform as follows:

$$\mathbb{E}_X\left[-\frac{(X-\mu_*)^2}{2}+\frac{(X-\mu)^2}{2}\right]=(\mu_*-\mu)\mathbb{E}_X\left[X-\frac{\mu_*+\mu}{2}\right]=-(\mu_*-\mu)\frac{\mu_*+\mu}{2},$$

and

$$\mathbb{E}_X\left[\{-\frac{(X-\mu_*)^2}{2}+\frac{(X-\mu)^2}{2}\}^2\right]=(\mu_*-\mu)^2\mathbb{E}_X\left[X-\frac{\mu_*+\mu}{2}\right]^2$$

$$=(\mu_*-\mu)^2\left\{1+\left(\frac{\mu_*+\mu}{2}\right)^2\right\}.$$

Then, in Example 9, $\Theta_*=\{\pm1\}$, so it includes $\mu_*=1$, $\mu=-1$. Therefore, there does not exist a c satisfying (2.22). However, in Example 10, since $\Theta_*=\{1\}$ and considering the range of Θ, we can have $\mu_*+\mu\geq2$. Thus, there exists a c satisfying (2.22).

A similar transformation can be made in Example 11. Considering the t-distribution with degrees of freedom m, substituting $\mu_*=0$ gives

$$\log\frac{p(x|\mu_*)}{p(x|\mu)}=-\frac{(x-\mu_*)^2}{2}+\frac{(x-\mu)^2}{2}=-\mu(x-\frac{\mu}{2})$$

and the square mean and mean are

$$\mathbb{E}[\{-\mu(X - \frac{\mu}{2})\}^2] = \mathbb{V}[-\mu(X - \frac{\mu}{2})] + \mathbb{E}[-\mu(X - \frac{\mu}{2})]^2$$

$$= \mu^2 \mathbb{V}[X] + (\frac{\mu^2}{2})^2 = \mu^2(\frac{m}{m-2} + \frac{\mu^2}{4})$$

and $\mathbb{E}[-\mu(X - \mu/2)] = \mu^2/2$, respectively. Thus, such a c in (2.22) exists because

$$2(\frac{m}{m-2} + \frac{\mu^2}{4}) \le 2(\frac{m}{m-2} + \frac{1}{4})$$

regardless of the value of μ. The fact that Examples 12 and 13 have relatively finite variance is derived from Proposition 2. ∎

In this book, we assume relatively finite variance and proceed with the discussion. This assumption implies that the optimal parameter set Θ is homogeneous. However, whether the distribution q is regular or not, and whether it is realizable or not, for the statistical model $\{p(\cdot|\theta)\}_{\theta \in \Theta}$ varies in individual cases.

In practice, though, we often deal with cases that are not regular but realizable. It might be helpful to understand this as a generalization that can handle such cases (problems like Example 10 are rare in practice).

2.4 Exponential Family

Assuming $J \ge 1$, if the conditional probability of $x \in \mathcal{X}$ under $\theta \in \Theta$ can be written using some $u : \mathcal{X} \to \mathbb{R}$, $v : \Theta \to \mathbb{R}^J$, $w : \mathcal{X} \to \mathbb{R}^J$ as

$$p(x|\theta) = u(x) \exp v(\theta)^\top w(x), \tag{2.23}$$

then this distribution is said to belong to the *exponential family*. Furthermore, if the distribution of the parameter $\theta \in \Theta$ depends on some *hyperparameter* $\phi \in \mathbb{R}^J$, and can be written as

$$\varphi(\theta|\phi) := \frac{\exp v(\theta)^\top \phi}{z(\phi)} \tag{2.24}$$

$$z(\phi) = \int_\Theta \exp(v(\theta)^\top \phi) d\theta$$

then, $\varphi(\theta|\phi)$ is said to be a *conjugate prior distribution*. Note that we use the same function v in (2.23) and (2.24).

Example 16 In the case of the normal distribution (2.6) from Example 3, we have
$J = 4$, $u(x) = \dfrac{1}{\sqrt{2\pi}}$, $v(\theta)^\top = [\dfrac{1}{\sigma^2}, \dfrac{\mu}{\sigma^2}, \dfrac{\mu^2}{\sigma^2}, \log \sigma^2]$, and $w(x)^\top = [-\dfrac{x^2}{2}, x, -\dfrac{1}{2}, -\dfrac{1}{2}]$ (note that u, v, w do not uniquely decompose). ∎

Here, let's define $\phi_n := \phi + \sum_{i=1}^{n} w(x_i)$. Then, we can transform the following expression:

$$
\varphi(\theta|\phi) \prod_{i=1}^{n} p(x_i|\theta) = \frac{\exp v(\theta)^\top \phi}{z(\phi)} \prod_{i=1}^{n} [u(x_i) \exp v(\theta)^\top w(x_i)]
$$

$$
= \exp v(\theta)^\top (\phi + \sum_{i=1}^{n} w(x_i)) \cdot \frac{1}{z(\phi)} \prod_{i=1}^{n} u(x_i)
$$

$$
= \frac{\exp v(\theta)^\top \phi_n}{z(\phi_n)} \cdot \frac{z(\phi_n)}{z(\phi)} \prod_{i=1}^{n} u(x_i). \tag{2.25}
$$

It is important to note that regardless of $\phi \in \mathbb{R}^J$, the integral with respect to θ in (2.24) equals 1, so the first half of (2.25) forms the posterior distribution, and the second half forms the marginal likelihood.

$$
p(\theta|x_1, \ldots, x_n) = \frac{\exp v(\theta)^\top \phi_n}{z(\phi_n)}
$$

$$
Z(x_1, \ldots, x_n) = \frac{z(\phi_n)}{z(\phi)} \prod_{i=1}^{n} u(x_i).
$$

The predictive probability is given by

$$
r(x|x_1, \ldots, x_n) = \frac{Z(x_1, \ldots, x_n, x)}{Z(x_1, \ldots, x_n)} = \frac{\dfrac{z(\phi_n + w(x))}{z(\phi)} \prod_{i=1}^{n} u(x_i) \cdot u(x)}{\dfrac{z(\phi_n)}{z(\phi)} \prod_{i=1}^{n} u(x_i)}
$$

$$
= u(x) \frac{z(\phi_n + w(x))}{z(\phi_n)}.
$$

Example 17 The prior distribution (2.9) can be interpreted as a conjugate prior with $J = 2$, $u(x) = \dfrac{1}{\sqrt{2\pi}} \exp(-\dfrac{x^2}{2})$, $v(\mu)^\top = [\mu, -\dfrac{\mu^2}{2}]$, $w(x)^\top = [x, 1]$, and $\phi^\top = [\phi_1, \phi_2] = [0, 1]$. In fact, for a general $[\phi_1, \phi_2]$,

$$z([\,\phi_1\ \phi_2\,])^T = \int_\Theta \exp\{\phi_1\mu - \frac{\phi_2\mu^2}{2}\}d\mu = \int_\Theta \exp\{-\frac{\phi_2}{2}(\mu - \frac{\phi_1}{\phi_2})^2 + \frac{\phi_1^2}{2\phi_2}\}d\mu$$

$$= \sqrt{\frac{2\pi}{\phi_2}}\,\exp(\frac{\phi_1^2}{2\phi_2})$$

can be obtained, thus $z(\phi) = z([0, 1]^T) = \sqrt{2\pi}$. Also,

$$\phi_n = \phi + \sum_{i=1}^n w(x_i) = [0, 1]^T + \sum_{i=1}^n [x_i, 1]^T = [\sum_{i=1}^n x_i, n + 1]^T$$

from which we can derive

$$z(\phi_n) = z([\sum_{i=1}^n x_i, n + 1]^T) = \sqrt{\frac{2\pi}{n + 1}}\,\exp\left\{\frac{(\sum_{i=1}^n x_i)^2}{2(n + 1)}\right\}.$$

From this, it becomes clear that (2.10), (2.11), and (2.12) hold. ∎

Example 18 Example 17 also holds when $J = 3$, $u(x) = \dfrac{1}{\sqrt{2\pi}}$, $v(\mu)^T = [-\dfrac{1}{2}$, $\mu, -\dfrac{\mu^2}{2}]$, $w(x)^T = [x^2, x, 1]$, and $\phi^T = [0, 0, 1]$. Furthermore, if we let $\phi^T = [\phi^2, \phi, 1]$, we can obtain the predictive distribution (2.18) when the prior distribution is set as (2.17). ∎

Example 19 The prior distribution (2.19) is a conjugate prior distribution with $J = 2$, $u(x) = 1$, $v(\theta)^T = [\log\theta, \log(1 - \theta)]$, $w(x)^T = [x, 1 - x]$, and $\phi^T = [\phi_1, \phi_2] = [a - 1, b - 1]$. In fact, for a general $[\phi_1, \phi_2]^T$,

$$z([\phi_1, \phi_2]^T) = \int_\Theta \exp\{\phi_1\log\theta + \phi_2\log(1 - \theta)\}d\theta = \int_\Theta \theta^{\phi_1}(1 - \theta)^{\phi_2}d\theta$$

can be obtained. Let the right-hand side be $B(\phi_1 + 1, \phi_2 + 1)$, then $z(\phi) = z([a - 1, b - 1]^T) = B(a, b)$. Also,

$$\phi_n = \phi + \sum_{i=1}^n w(x_i) = [a - 1, b - 1]^T + \sum_{i=1}^n [x_i, 1 - x_i]^T = [k + a - 1, n - k + b - 1]^T$$

from which we can derive

$$z(\phi_n) = z([k + a - 1, n - k + b - 1]^T) = B(k + a, n - k + b).$$

This value is the marginal likelihood $Z(x_1, \ldots, x_n)$, and the posterior and predictive distributions are given by

$$p(\theta|x_1, \ldots, x_n) = \frac{\theta^{k+a-1}(1-\theta)^{n-k+b-1}}{B(k+a, n-k+b)}$$

and (2.20), respectively. Furthermore, if we set $a = b = 1$, we get (2.14), (2.15), and (2.16). In fact,

$$C_{n,k} := B(k+1, n-k+1) = \int_0^1 (\frac{\theta^{k+1}}{k+1})'(1-\theta)^{n-k}d\theta$$

$$= \left[\frac{\theta^{k+1}}{k+1}(1-\theta)^{n-k}\right]_0^1 + \frac{n-k}{k+1}\int_0^1 \theta^{k+1}(1-\theta)^{n-k-1} = \frac{n-k}{k+1}C_{n,k+1}$$

and from $C_{n,n} = (n+1)^{-1}$, we get

$$C_{n,k} = \frac{n-k}{k+1}C_{n,k+1} = \cdots = \frac{n-k}{k+1}\cdots\frac{1}{n}C_{n,n} = \frac{(n-k)!k!}{n!} \cdot \frac{1}{n+1} = \frac{(n-k)!k!}{(n+1)!}$$

holds true.   ∎

Exercises 1–13

1. Using (2.3), prove (2.4) and (2.5).
2. Show that for the true distribution q in (2.2), the mean of an odd function $f :$ $(-\infty, \infty) \to \mathbb{R}$, $\int_{-\infty}^{\infty} f(x)q(x-\mu)dx$ is 0, and using (2.1) as well as

$$\int_{-\infty}^{\infty} \frac{z^2}{\sqrt{2\pi}} \exp(-\frac{z^2}{2})dz = 2\int_0^{\infty} \frac{z}{\sqrt{2\pi}}\{-\exp(-\frac{z^2}{2})\}'dz$$

$$= 2\left\{\left[-\frac{z}{\sqrt{2\pi}}\exp(-\frac{z^2}{2})\right]_0^{\infty} + \int 0^{\infty}\frac{1}{\sqrt{2\pi}}\exp(-\frac{z^2}{2})dz\right\} = 1$$

prove that $\mathbb{E}[X] = \mu$, $\mathbb{V}[X] = \sigma^2$, $\mathbb{E}[(X-\mu)^4] = 3\sigma^4$.
3. In Example 4's statistical model, when the prior distribution is (2.9), show that $\varphi(\theta) \geq 0$ and $\int_{-\infty}^{\infty} \varphi(\theta)d\theta = 1$. Also, derive the marginal likelihood (2.10), posterior distribution (2.11), and predictive distribution (2.12).
4. In Example 5's statistical model, let there be k instances of 1 in $x_1, \ldots, x_n \in \mathcal{X}$. When the prior distribution is (2.13), show that $\varphi(\theta) \geq 0$ and $\int_0^1 \varphi(\theta)d\theta = 1$. Also, derive the marginal likelihood (2.14), posterior distribution (2.15), and predictive distribution (2.16).
5. Fill in the blanks below to execute the graph of the posterior distribution in Fig. 2.1 with hyperparameters $\phi = -3, -1, 1, 3$, and the first $n = 1, 5, 10$ samples. Verify that it produces the desired graph.

```
1   f <- function(x,mu,sig2) (2*pi*sig2)^(-1/2)*exp(-(x-mu)^2/2)
2                           ## Definition of normal distribution
3   phi.seq <- # Blank (1) #
4   n.seq <-   # Blank (2) #
5   m <- length(phi.seq)
6   l <- length(n.seq)
7   n <- 30                 ## Number of samples n
8   x <- rnorm(n)           ## Generate n normal random numbers
9   par(mfrow = c(2,2))     ## Generate 4 graphs
10  for(phi in phi.seq){
11     plot(0, 0, xlim=c(-5,7), ylim=c(0,0.5), type="n")
12     for(k in 1:l){
13        nn <- n.seq[k]      ## First n samples
14        mu <-     # Blank (3) #
15        sig2 <-  # Blank (4) #
16        curve(f(x,mu,sig2), col=k+1, add=TRUE)  ## Draw the curve
17        title(paste("phi=",phi))
18     }
19     ## Draw the curve of the true distribution
20     curve(dnorm(x), lwd=2, lty=2, col=1, add=TRUE)
21     legend("topright",c("True", n.seq), lty=c(2, rep(1,4)),
22        lwd=c(2,rep(1,4)), col=1:(1+1))
23  }
24  par(mfrow = c(1,1))
```

6. In R language, generate a graph similar to Fig. 2.4.

7. Show that in Example 9, the true distribution and statistical model are not homogeneous, not realizable, not regular, and do not have a relatively finite variance.

8. Show that in Example 10, the true distribution and statistical model are homogeneous, not realizable, not regular, and have a relatively finite variance.

9. Show that in Example 11, the true distribution and statistical model are homogeneous, not realizable, regular, and have a relatively finite variance.

10. Show that in Example 12, the true distribution and statistical model are homogeneous, realizable, not regular, and have a relatively finite variance. However, you may use Proposition 2.

11. Show that in Example 13, the true distribution and statistical model are homogeneous, realizable, regular, and have a relatively finite variance. However, you may use Proposition 2.

12. In Example 14, show that when $a_* b_* = 0$, there are multiple (a, b) pairs for which $K(a, b) = 0$.

13. In Example 19, for non-negative integers l, m, express the value of $B(l, m)$ using factorials!.

The abbreviated solutions for each chapter's exercise problems can be found at https://bayesnet.org/books.

Chapter 3
MCMC and Stan

In Bayesian statistics, it is generally difficult to mathematically derive the posterior distribution, except in special cases. Instead, it is common to generate random numbers following the posterior distribution and perform integration calculations based on their frequency. In this chapter, we will discuss Markov Chain Monte Carlo (MCMC) methods, which generate random numbers following the posterior distribution using Markov chains. Bayesian theory by Watanabe seeks to obtain asymptotic posterior distributions in general situations without assuming regularity. Since MCMC generates random numbers without assuming regularity, it is an effective means of experimentally verifying the results of Watanabe's Bayesian theory. In fact, MCMC is also used to calculate WAIC and WBIC introduced in Chap. 6 and beyond. In particular, this chapter introduces two types of MCMC, the Metropolis-Hastings method and the Hamiltonian method, to understand their principles. Stan, which we will cover in this book, is a realization method for the latter category of MCMC. Since we will also cover Stan in the following chapters, we will limit our discussion in this chapter to the minimum necessary understanding of how to describe Stan files and execute programs.

3.1 MCMC and Metropolis-Hastings Method

Given a statistical model $p(\cdot|\theta)$, $\theta \in \Theta \subseteq \mathbb{R}^d$, samples $x_1, \ldots, x_n \in \mathcal{X}$, and a prior distribution $\varphi(\cdot)$, we consider the calculation of the expected value of a function $f : \Theta \to \mathbb{R}$ with respect to the posterior distribution $p(\theta|x_1, \ldots, x_n)$, which is

$$\int_\Theta f(\theta) p(\theta|x_1, \ldots, x_n) d\theta.$$

J. Suzuki, *WAIC and WBIC with R Stan*,
https://doi.org/10.1007/978-981-99-3838-4_3

Example 20 Let $f(\theta) = p(x|\theta)$, and let us seek the value of the predictive distribution, $r(x|x_1, \ldots, x_n)$, for each $x \in \mathcal{X}$, given by

$$r(x|x_1, \ldots, x_n) = \int_\Theta p(x|\theta) p(\theta|x_1, \ldots, x_n) d\theta.$$
∎

In addition, WAIC and WBIC, which are also discussed in this book, can be expressed in this form by appropriately setting the function f.

In the previous chapter, we confirmed that given a statistical model $\{p(\cdot|\theta)\}_{\theta \in \Theta}$ and a prior distribution $\varphi(\cdot)$, the posterior distribution $p(\theta|x_1, \ldots, x_n)$ and the predictive distribution $r(x|x_1, \ldots, x_n)$ for $x \in \mathcal{X}$ can be determined from a sample $x_1, \ldots, x_n \in \mathcal{X}$, provided $\theta \in \Theta$. However, in cases such as when the conjugate prior distribution of exponential family is unknown (see Examples 4 and 5), the posterior and predictive distributions cannot necessarily be solved analytically. In such cases, numerical computation is usually necessary.

We generate random variables $\theta_1, \ldots, \theta_K$ following the posterior distribution $p(\theta|x_1, \ldots, x_n)$ with $\theta \in \Theta$, and approximate the integral as

$$\frac{1}{K} \sum_{k=1}^{K} f(\theta_k), \quad \theta_k \sim p(\cdot|x_1, \ldots, x_n). \tag{3.1}$$

Especially, the method of generating random variables for the stationary distribution of a Markov chain is called the *Markov Chain Monte Carlo (MCMC)* method. Here, a sequence of random variables $\theta_1, \theta_2, \ldots$ that only depend on θ_k and are independent of $\theta_1, \ldots, \theta_{k-1}$ is called a *Markov chain*, and its conditional probability is denoted as $P(\theta_{k+1}|\theta_k)$ $(k = 1, \ldots, K-1)$. In this case, θ_{k+1} is said to be conditionally independent of $\theta_1, \ldots, \theta_{k-1}$ given θ_k.

Then, if the following two conditions hold, it is known that the accuracy of Equation (3.1) improves arbitrarily as K increases.

1. For any $\theta, \theta' \in \Theta$,

$$P(\theta|\theta') p(\theta'|x_1, \ldots, x_n) = P(\theta'|\theta) p(\theta|x_1, \ldots, x_n). \tag{3.2}$$

2. The probability of reaching any neighborhood of $\theta \in \Theta$ is not zero.

Equation (3.2) is called the *detailed balance condition*, and the second condition is called *ergodicity*.

The detailed balance condition pertains to the $P(\cdot|\cdot)$ probability distribution (referred to as the transition probability of the Markov chain) that is set up in MCMC. It is a condition that ensures that the stationary probability is equal to $p(\theta|x_1, \ldots, x_n)$, the probability distribution of θ given the observed data $x_1, \ldots, x_n$, even if θ transitions to other values through (3.2) and there are transitions from other values to θ.

The Metropolis-Hastings (MH) algorithm is a representative method of MCMC. First, we prepare a symmetric conditional probability,

$$s(\theta'|\theta) = s(\theta|\theta') \, , \ \theta, \theta' \in \Theta \tag{3.3}$$

with

$$\int_\Theta s(\theta'|\theta)d\theta' = 1 \, ,$$

where $s(\theta'|\theta) \geq 0, \theta, \theta' \in \Theta$, and select an initial value θ_1. Then, defining a function $H(\theta)$ on the set Θ as

$$H(\theta) := -\sum_{i=1}^{n} \log p(x_i|\theta) - \log \varphi(\theta) \, ,$$

we execute the following steps for $k = 1, 2, \ldots,$

1. Generate $\theta \in \Theta$ according to the probability $s(\theta|\theta_k)$.
2. Obtain θ_{k+1} as follows: with probability $Q_k := \min\{1, \exp(-\{H(\theta) - H(\theta_k)\})\}$, set $\theta_{k+1} \leftarrow \theta$ (accepted), otherwise set $\theta_{k+1} \leftarrow \theta_k$ (rejected).

Here, $s(\theta'|\theta), \theta, \theta' \in \Theta$ is called the proposal distribution, and the resulting $P(\theta'|\theta)$, $\theta, \theta' \in \Theta$ becomes the transition probability. The acceptance and rejection of the proposal correspond to the acceptance and rejection of the transition in Step 2, respectively.

The following proposition holds for the Markov chain generated by the MH algorithm, provided that $s(\cdot|\cdot)$ is symmetric (3.3):

Proposition 3 *The detailed balance condition (3.2), i.e.,*

$$P(\theta'|\theta) \exp\{-H(\theta)\} = P(\theta|\theta') \exp\{-H(\theta')\}$$

holds for the MH algorithm.

Proof Please refer to the appendix at the end of the chapter.　∎

For some time after execution, there is an influence of the initial values and the stationary distribution of the Markov chain is not reached. The period until the influence of the initial values disappears is called *burn-in*. Sampling during this period is not used for approximating the distribution.

Example 21 We use the Metropolis-Hastings algorithm to generate random numbers for σ^2 following the posterior distribution $p(\mu, \sigma^2|x_1, \ldots, x_n)$ of a normal distribution with known mean μ and unknown variance σ^2, based on a sample of $n = 30000$ data points $x_1, \ldots, x_n$. Assuming $\mu = 10$ and $\sigma = 3$, we generate n random numbers and assume that the arithmetic mean $\hat{\mu}$ is the correct value for μ. We then sample 1000 data points from the original dataset and display them as a histogram (Fig. 3.1):

Fig. 3.1 Artificial data used in Example 21. We generated $n = 30000$ data points with $\mu = 10$ and $\sigma = 3$, and randomly sampled 1000 of them

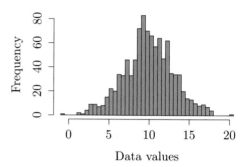

```
1   pop <- rnorm(30000,10,3)
2   obs <- pop[as.integer(runif(1000, min = 1, max = 30001))]
3   mu_obs <- mean(obs)
4   hist(obs, breaks=35, xlab="Data values", ylab="Frequency",
5       main="Data used in the example",col=3)
```

As a transition rule, we set the rule $\sigma \leftarrow N(\sigma, 0.5^2)$, which adds a random number from $N(0, 0.5^2)$ to the current value $\theta[2]$ of σ, without changing the value $\theta[1]$ of μ (function trans). This satisfies condition (3.3):

```
1   trans<- function(theta) c(theta[1],abs(rnorm(1, theta[2],0.5)))
```

We take the absolute value to ensure that the updated value of σ does not become negative.[1] We also set the function lik to calculate the likelihood $(- \sum_{i=1}^{n} \log p(x_i|\theta))$, and the acceptance rule function accept, which determines whether or not to accept and with what probability (Step 2). We should use $H(\theta)$ instead of $- \sum_{i=1}^{n} \log p(x_i|\theta)$ for the lik. However, for simplicity, we remove the effect of the prior distribution in Sects. 3.1 and 3.2:

```
1   lik <- function(theta,data) sum(-log(theta[2]*sqrt(2*pi))-((data-theta
2       [1])**2)/(2*theta[2]**2))
3   ## data is sample x_1, ... , x_n (vector of length n)
```

```
1   accept <- function(x, y) runif(1,0,1) < exp(y-x)
```

Since runif(1,0,1) is the uniform distribution on $[0, 1]$, if $y > x$, TRUE is returned unconditionally. Then, we construct the function metropolis_hastings to calculate the likelihood of the proposed parameters and the new and old parameters. Since we assume the same prior distribution, it follows that

---

[1] Therefore, (3.3) does not strictly hold.

$$H(\theta) - H(\theta') = -\sum_{i=1}^{n} \log p(x_i|\theta) + \sum_{i=1}^{n} \log p(x_i|\theta').$$

```
metropolis_hastings <- function(theta,data){
    theta_new <- trans(theta)
    H_old <- lik(theta,data)
    H_new <- lik(theta_new,data)
    return(list(theta=theta_new, old=H_old, new=H_new))
}
```

The following is the function MCMC to execute the MCMC. $\mu = \hat{\mu}$ is fixed and does not change. The values of the parameters (μ, σ^2) are stored in theta[1], theta[2]. The column of theta[2] is the random number sequence of σ following the posterior distribution we are trying to obtain.

```
MCMC <- function(proc, theta_init,iters,data){
    theta <- theta_init          ## initial value of theta
    output <- NULL
    accept_reject <- NULL
    for(i in 1:iters){
        res <- proc(theta,data)      ## proc is metropolis_hastings
        if (accept(res$old,res$new)){  ## compare old and new to accept/
          reject
            theta <- res$theta     ## proposed theta
            accept_reject <- c(accept_reject,1)
        }
        else {
            accept_reject <- c(accept_reject,0)
        }
        output <- c(output,res$theta)
    }
    return(list(output=output, accept_reject=accept_reject))
}
```

Figure 3.2 is obtained sequentially by executing the following:

```
m <- 50000;
# m <- 100      To run with m set to 100, uncomment the line
result = MCMC(metropolis_hastings, c(mu_obs,3), m, obs)
output <- result$output
output2 <- output[seq(2,2*m,2)]
colors <- 2*result$accept_reject+2
pchs <- -3*result$accept_reject+4
plot(1:m, output2[1:m], col=colors[1:m], xlab="Number of iterations",
    ylab="Value of sigma",ylim=c(1.0,5.0), main="Generated parameters")
legend("bottomright",legend=c("accepted","rejected"),col=c(4,2), pch=c
        (1,1))
```

```
hist(output2,breaks=c(0,seq(1,5,0.01)),xlim=c(1,5),col=4,
    xlab="Value of sigma", ylab="Probability density", main="Posterior
    distribution of sigma")
```

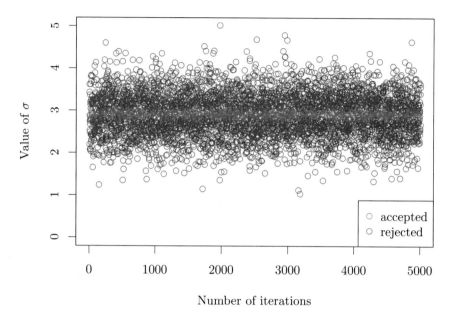

Fig. 3.2 Random numbers for σ following the posterior distribution. Values farther from the center are rejected (in red)

```
1  sigma <- sqrt(sum(output2^2)/length(output2))
2  hist(obs, breaks=35, xlab="Data values", ylab="Probability density",
3          main="Original data and predictive distribution",
4      freq=FALSE,col=3)
5  curve(dnorm(x,mu_obs,sigma),mu_obs-3*sigma,mu_obs+3*sigma, add=TRUE,col
6      =2,lwd=2)
```

■

Also, regardless of whether performance guarantees can be obtained or not, there is a strategy called thinning, in which not all of the series of posterior distributions of parameters are used, but every few values like θ_{2k} and θ_{3k+1} are taken. The aim is to reduce the autocorrelation between adjacent values in the series. Furthermore, there is a method to extract multiple series and check if there are any convergence problems to ensure the quality of MCMC solutions. Each series in this case is called a chain. The Metropolis-Hastings method is considered difficult to choose the optimal method for setting the updating rule. For example, in Example 21, the updating rule uses $N(\sigma, 0.5^2)$ to select the next σ, but there is no basis for the standard deviation of 0.5 to be valid. If this standard deviation is too small, the risk of strong correlation between the series and the lack of ergodicity arises.

Posterior distribution of σ

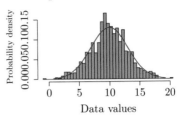

Fig. 3.3 Random numbers of σ following the posterior distribution. It can be seen that the shape is close to a normal distribution. As discussed in Chap. 5, for regular models with large sample sizes, it is known that the posterior distribution of parameters takes a shape close to a normal distribution (left). Using the obtained posterior distribution of σ, we calculated the predictive distribution of future $x \in \mathcal{X}$ and overlaid it on the histogram of the original data in Fig. 3.1 (right). It appears that noise has been removed

3.2 Hamiltonian Monte Carlo Method

In Stan, which is actively used in this book, the *Hamiltonian Monte Carlo method (HMC)* is adopted instead of the Metropolis-Hastings method. Also, updates like $\sigma \leftarrow N(\sigma, 0.5^2)$ are called random walks in the context of Markov chains. By applying Hamilton's equations, a kind of Newton's equations of motion, without relying on random walks, it is possible to dramatically change the state while maintaining a low rejection rate.

Let $\theta(t) = [\theta_1(t), \ldots, \theta_d(t)]^\top$, $p(t) = [p_1(t), \ldots, p_d(t)]^\top$ be the positions and momenta (product of mass and velocity) of d particles at time t, respectively. The sum of kinetic energy $V(p)$ and potential energy $U(\theta)$,

$$H(p, \theta) = V(p) + U(\theta)$$

is called the Hamiltonian. Furthermore,

$$\frac{d\theta(t)}{dt} = \nabla_p V(p) \tag{3.4}$$

and

$$\frac{dp(t)}{dt} = -\nabla_\theta U(\theta) \tag{3.5}$$

are called Hamilton's equations. In terms of components, each is $\dfrac{d\theta_j(t)}{dt} = \dfrac{\partial V(p)}{\partial p_j}$, $\dfrac{dp_j(t)}{dt} = -\dfrac{\partial U(\theta)}{\partial \theta_j}$, $j = 1, \ldots, d$. In general, if $u = \phi(t)$, $v = \psi(t)$ are differentiable, and $f(u, v)$ is totally differentiable[2] with respect to u, v, the derivative of the composite function $w = f(u, v) = f(\phi(t), \psi(t))$ can be written as follows:

$$\frac{dw}{dt} = \frac{\partial f}{\partial u}\frac{du}{dt} + \frac{\partial f}{\partial v}\frac{dv}{dt}.$$

Using this, the energy conservation law

$$\begin{aligned}
\frac{dH(p, \theta)}{dt} &= \nabla_p H(p, \theta)^\top \frac{dp(t)}{dt} + \nabla_\theta H(p, \theta)^\top \frac{d\theta(t)}{dt} \\
&= \nabla_p V(p)^\top \frac{dp(t)}{dt} + \nabla_\theta U(\theta)^\top \frac{d\theta(t)}{dt} \\
&= \sum_{j=1}^d \frac{\partial V(p)}{\partial p_j}\frac{dp_j(t)}{dt} + \sum_{j=1}^d \frac{\partial U(\theta)}{\partial \theta_j}\frac{d\theta_j(t)}{dt} \\
&= \sum_{j=1}^d \frac{\partial V(p)}{\partial p_j}(-\frac{\partial U(\theta)}{\partial \theta_j}) + \sum_{j=1}^d \frac{\partial U(\theta)}{\partial \theta_j}\frac{\partial V(p)}{\partial p_j} = 0
\end{aligned}$$

is satisfied (Exercise 17).

In the following, let

$$V(p) := \frac{1}{2}\|p\|^2$$

and

$$U(\theta) := -\sum_{i=1}^n \log p(x_i | \theta) - \log \varphi(\theta),$$

where $\|p\|^2$ is the sum of the squares of the components of p. Then, the MCMC algorithm is constructed as follows. Determine the initial value of θ and repeat the following:

1. Generate momentum $p \sim N(0, 1)$.
2. Calculate $H(p, \theta)$.
3. Obtain $(p(T), \theta(T)) = (p', \theta')$ after a fixed time T has elapsed from $(p(0), \theta(0)) = (p, \theta)$ based on the Hamilton equations.
4. Accept θ' with probability $\min\{1, \exp\{H(p, \theta) - H(p', \theta')\}\}$.

In the following, we will discuss the method to obtain the trajectory from $(p(0), \theta(0)) \to (p(T), \theta(T))$, and the reasons why $H(p, \theta)$ and $H(p', \theta')$ do not match.

The trajectory is obtained by updating (p, θ) at discrete times. By taking $\epsilon > 0$ and updating (3.4) (3.5) with

---

[2] $f : \mathbb{R}^2 \to \mathbb{R}$ is totally differentiable at $(x, y) \in \mathbb{R}^2$ if there exist $A, B \in \mathbb{R}$ such that $\frac{f(x+h, y+k) - f(x, y) - (Ah + Bk)}{\sqrt{h^2 + k^2}} \to 0 \ ((h, k) \to (0, 0))$.

$$p(t + \epsilon) \leftarrow p(t) + \epsilon \cdot \frac{dp}{dt} = p(t) - \epsilon \cdot \nabla_\theta U(\theta)$$

$$\theta(t + \epsilon) \leftarrow \theta(t) + \epsilon \cdot \frac{d\theta}{dt} = \theta(t) + \epsilon \cdot \nabla_p V(p)$$

(*Euler method*), the trajectory of (p, θ) can be obtained. However, in the Euler method, errors accumulate due to discretization, so the following update procedure is used:

1. $p(t + \epsilon/2) \leftarrow p(t) - \frac{\epsilon}{2} \cdot \nabla_\theta U(\theta)$.
2. $\theta(t + \epsilon) \leftarrow \theta(t) + \epsilon \cdot p(t + \epsilon/2)$.
3. $p(t + \epsilon) \leftarrow p(t + \epsilon/2) - \frac{\epsilon}{2} \cdot \nabla_\theta U(\theta)$

(*Leapfrog method*). Even with the Leapfrog method, there is an error, and the values of $H(p, \theta)$ and $H(p', \theta')$ are slightly different. However, since both are almost the same, the probability of acceptance is high.

Also, the momentum p is changed in each cycle. This means that each cycle traces a trajectory with different energy $H(p, \theta)$. The space represented by $(p(t), \theta(t))$ is called the phase space, and each trajectory is represented by contour lines (Fig. 3.4).

Example 22 Consider the simplest case:

$$H(p, \theta) = U(\theta) + V(p) = \frac{\theta^2}{2} + \frac{p^2}{2}.$$

From (3.4), (3.5), $\frac{d\theta}{dt} = p$, $\frac{dp}{dt} = -\theta$, so for $a, r \in \mathbb{R}$, the general solution is $\theta(t) = r \cos(a + t)$, $p(t) = -r \sin(a + t)$. Under the initial conditions $\theta(0) = 1$, $p(0) = 0$, we have $a = 0$ and $r = 1$. To find this solution, we observed the behavior using the following R language code. For both $\epsilon = 0.1$ and $\epsilon = 0.3$, the $(p(t), \theta(t))$ of

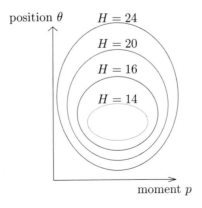

Fig. 3.4 Phase space. The contour lines are represented by (p, θ) with equal energy

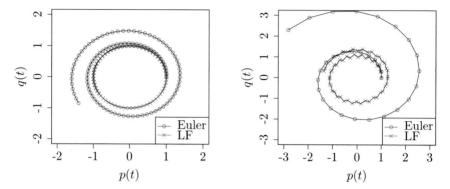

Fig. 3.5 $\epsilon = 0.1$ (left) and $\epsilon = 0.3$ (right). In the case of the Euler method, since the value of the derivative one time step before is used, errors tend to accumulate, such as when the trajectory draws a circle. On the other hand, such problems are less likely to occur in the Leapfrog method

the Euler and Leapfrog methods were plotted using the R language code below. In the case of the Euler method, since the value of the derivative one time step before is used, errors tend to accumulate, such as when the trajectory draws a circle (Fig. 3.5). This property becomes more evident when the value of ϵ is large: ∎

```
L <- 2
M <- 100
eps <- 0.1
# You can also run with L <- 3; M <- 30; eps <- 0.3
```

```
euler <- function(p,q){
  r <- p-eps*q
  s <- q+eps*p
  return(list(p=r,q=s))
}
```

```
leapfrog <- function(p,q){
  p <- p-eps/2*q
  q <- q+eps*p
  p <- p-eps/2*q
  return(list(p=p,q=q))
}
```

```
draw <- function(proc,pch,col,P=0,Q=1){
  p <- rep(0,M)
  q <- rep(0,M)
  p[1] <- P
  q[1] <- Q
  for(i in 1:(M-1)){
    res <- proc(p[i],q[i])
    p[i+1] <- res$p
    q[i+1] <- res$q
  }
  points(p,q,pch=pch,col=col)
```

```
12    lines(p,q)
13    }
```

```
1    plot(0,xlim=c(-L,L),ylim=c(-L,L))
2    draw(euler,1,2)
3    draw(leapfrog,4,4)
4    legend("bottomright",legend=c("Euler","Leapfrog"),lwd=1,pch=c(1,4),col=c
         (2,4))
```

In HMC, the following two general-purpose functions are needed. First, the function `Leapfrog` below is an extension of the already defined `leapfrog` for the general case:

```
1    eps <- 0.01
2    Leapfrog <- function(U,p,theta){
3       p <- p - eps/2 * grad_U(U,theta)
4       theta <- theta + eps * p
5       p <- p + eps/2 * grad_U(U,theta)
6       return(list(p=p,theta=theta))
7    }
```

The function `Leapfrog` requires a function to numerically differentiate. For example, it can be constructed as follows:

```
1    grad_U <- function(U,theta){
2       p <- length(theta)
3       h <- 0.01
4       f <- U(theta)
5       diff <- NULL
6       for(i in 1:p){
7          theta_h <- theta
8          theta_h[i] <- theta[i]+h
9          diff[i] <- (U(theta_h)-f)/h
10      }
11      return(diff)
12   }
```

Example 23 In Example 21, we applied the `metropolis_hastings` function to MCMC. In this section, we apply the `hamiltonian` function to the same data. In this example, we find a sequence of random numbers following the posterior distribution of (μ, σ) under the samples $x_1, \ldots, x_n$:

```
1    L <- 20
2    hamiltonian <- function(theta,data){
3       U <- function(theta)lik(theta,data)
4       p <- rnorm(length(theta))
5       H_old <- U(theta)-sum(p^2)/2
6       for(i in 1:L){
7          res <- Leapfrog(U,p,theta)
8          p <- res$p
9          theta <- res$theta
10      }
11      H_new <- U(theta)-sum(p^2)/2
```

```
12      return(list(theta=theta, old=H_old, new=H_new))
13   }
```

Using the HMC method, we generated 10,000 random numbers following the posterior distribution of (μ, σ) and produced the corresponding histograms (Fig. 3.6). To obtain these, we executed the following code:

```
1   m <- 10000
2   result <- MCMC(hamiltonian,c(mu_obs,3), m, obs)
3   output <- result$output
4   output1 <- output[seq(1,2*m,2)]
5   output2 <- output[seq(2,2*m,2)]
6   s=length(output1)
7   plot(1:s, output1, xlab="Number of iterations", ylab="Value of mu",
8       ylim=c(9.5,10.5),type="l",col=2)
9   plot(1:s, output2, xlab="Number of iterations", ylab="Value of sigma",
10      ylim=c(2.75,3.50),type="l",col=4)
11  hist(output1,breaks=c(0,seq(9,11,0.02)),xlim=c(9,11),col=2,
12          xlab="Value of mu", ylab="Probability density", main="
        Posterior distribution of mu")
13  hist(output2,breaks=c(0,seq(2.5,3.75,0.01)),xlim=c(2.75,3.5),col=4,
14          xlab="Value of sigma", ylab="Probability density", main="
        Posterior distribution of sigma")
```

■

Each time the above functions are executed, the initial momentum p is determined randomly, and (p, θ) is updated while keeping the sum of the kinetic and potential energies constant, and the obtained θ' becomes the position of the stopped particle. In the program, U(theta), H_old, and H_new correspond to the potential energy and the total energy multiplied by -1, and when the process returns to MCMC, the accept function accepts with certainty if H_new is larger than H_old, and with high probability if they are close.

In *Stan*, a variant of HMC called *No U-turn Sampler (NUTS)* is used.

3.3   Stan in Practice

Stan is an implementation of MCMC as discussed in the previous sections. That is, it has the function of generating a large number of random numbers following the posterior distribution. Therefore, it is not used with existing information criteria such as AIC and BIC, but demonstrates its power in calculating WAIC and WBIC.

Stan is provided as the CRAN package "rstan". Create the Stan code, store it in a file named "***.stan" in the same folder as the R environment, and refer to it from the R language:

```
1   install.packages("rstan", dependencies = TRUE) # If not installed
2   library(rstan) # Every time you start the R environment
```

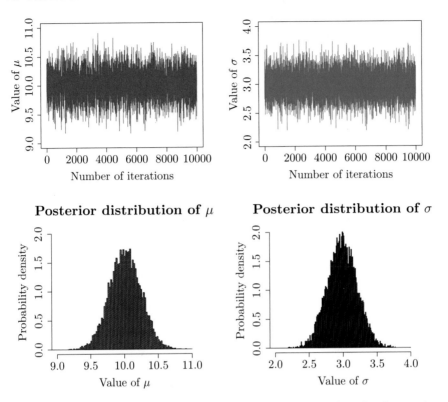

Fig. 3.6 Using the HMC method, we generated 10,000 random numbers following the posterior distribution of (μ, σ). We also produced the corresponding histograms (bottom two figures)

Be careful when installing Stan, as it can be tricky. It will not work if the versions of R, Rtools, and the `rstan` package are not appropriate (this book assumes versions R 4.2.3, Rtools 4.2, rstan 2.26.21).

Place the following binomial distribution, normal distribution, simple regression, multiple regression, and mixed normal distribution Stan codes (13 files from `model1.stan` to `model13.stan`) in the folder where R is executed, and then execute the following R code line by line and check the output results.

First, a Stan code consists of the following blocks. We will explain them in order through examples. In this book, Stan files are enclosed in orange frames to distinguish them from R code.

Block type	Role
Data	Definition of data and sample size used
Parameters	Specification of parameters for which posterior distribution is desired
Transformed parameters	Specification of parameter transformation
Model	Specification of model structure
Generated quantities	Specification when posterior distribution is desired separately from model specification

3.3.1 Binomial Distribution

Suppose we conduct a trial with a success probability of $0 < p < 1$ and observe D successes out of N attempts. We want to estimate the posterior distribution of p based on this fact. We constructed the Stan file as follows:

model1.stan

```
1  parameters{
2    real<lower=0,upper=1>p;    // The parameter for binomial distribution
3        0<p<1 is a real number, so real
4  }
5  model{
6    15 ~ binomial(30,p);       // 15 occurrences in 30 trials with
7        probability p
8  }
```

Put the parameter θ in the parameters block, and put the information about the prior distribution $\varphi(\theta)$ and the likelihood $p(y|\theta)$ in the model block. In this Stan code, the data y is fixed at $D = 15$ and $N = 30$. There are also the following rules when writing Stan code:

┌─ Rules for Writing Stan Code ──────────────────────────

  • Enclose blocks in curly braces
  • Put a semicolon at the end of each line
  • Comments are after // (double slash)
  • Put a blank line at the end of the file

Let's use this Stan file and execute it from R code:

```
1  fit1 <- stan("model1.stan")
2  # Check the results
3  # If all parameter Rhat values are 1.1 or lower, it is considered OK
4  fit1
5  # Trace of parameter sampling
6  stan_trace(fit1, pars="p")
7  # Posterior distribution of the parameter
8  stan_dens(fit1, pars="p")
9  # Density estimation of the posterior distribution of the parameter
```

```
10   # If it has converged, all chains should overlap nicely
11   stan_dens(fit1, pars="p", separate_chains = TRUE)
12   # If it converges to the stationary distribution, there is no correlation
13        with the previous sample
14   # High autocorrelation means it has not converged
15   stan_ac(fit1, pars="p", separate_chains = TRUE)
16   # Extract the element as p
17   # Since extract may exist in other packages, prepend rstan::
18   p <- rstan::extract(fit1)$p
19   # Look at the frequency distribution
20   hist(p)
21   mean(p)
```

Besides the output below, we obtain a histogram (Fig. 3.7):

....

```
> fit1
Inference for Stan model: anon_model.
4 chains, each with iter=2000; warmup=1000; thin=1;

post-warmup draws per chain=1000, total post-warmup draws=4000.
       mean se_mean   sd    2.5%    25%    50%    75%  97.5% n_eff Rhat
p      0.50    0.00 0.09    0.33   0.44   0.50   0.56   0.67  1391    1
lp__ -22.69    0.02 0.73  -24.71 -22.86 -22.41 -22.23 -22.18  1692    1

Samples were drawn using NUTS(diag_e) at Fri May 12 10:26:32 2023.
For each parameter, n_eff is a crude measure of effective sample size,
and Rhat is the potential scale reduction factor on split chains (at
convergence, Rhat=1).
```

....

```
> mean(p)
[1] 0.5007415
```

Fig. 3.7 The table obtained by the execution for a binomial distribution

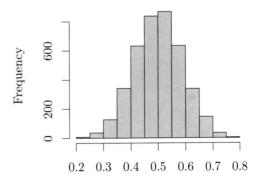

Histogram of p

The above Rhat represents the ratio of within-chain variance to between-chain variance, and if this value is within 1.1, it can be considered converged. Here, "15" and "30" can be used as input data:

model2.stan

```
1   data{
2     int D;  // Number of occurrences
3     int N;  // Number of trials
4   }
5   // The parameters block is the same
6   model{
7     D ~ binomial(N,p);
8   }
```

```
1   # Specify the data
2   data_list <- list(D=20, N=30)
3   fit2 <- stan("model2.stan", data=data_list)
4   stan_dens(fit2, pars="p")
5   # You can compile the model first
6   model2 <- stan_model("model2.stan")
7   fit2 <- sampling(model2, data=data_list)
8   # Add options to sampling
9   fit2 <- sampling(model2, data=data_list, iter=5000, warmup=1000, chains
10     =3, cores=3)
11  fit2
```

The options for sampling are as follows:

Option	Spec.	Default	Notes
Number of samples	iter=	2000	More than 2000 may be needed for complex models
Burn-in period	warmup=	iter/2	No need to touch
Thinning interval	thin=	1	thin=2 for every other one
Number of chains	chains=	4	At least 2 are needed to confirm convergence
Number of cores	cores=	1	Parallelization is possible, less than or equal to the number of Markov chains

If a prior distribution is not specified, it defaults to a uniform distribution. However, if the sample size is small, it is better to set an appropriate prior distribution for better estimation. As discussed in Chap. 2, the Beta distribution (2.19) is often used for binomial distributions. The uniform distribution can also be represented (just set $a = b = 1$):

model3.stan

```
1   // data, parameters blocks are the same
2   model{
3     p ~ beta(1,1);    // p's prior distribution is the same for Beta
         distribution with 1, -1 ~ uniform(0,1)
4     D ~ binomial(N,p);
5   }
```

Check how the results differ when specifying the prior distribution with model2.stan:

```
1  data_list <- list(D=20, N=30)
2  fit3 <- stan("model3.stan", data=data_list)
3  fit3
```

3.3.2 Normal Distribution

In the following, we want to estimate the posterior distribution of the mean and variance from 100 data points, assuming they follow a normal distribution. In the case of a normal distribution, the mean is often assumed to have a normal prior distribution, and the standard deviation is often assumed to have a half-Cauchy prior distribution (Cauchy distribution is symmetric about the origin, but only positive values are taken):

```
1  curve(dcauchy(x),0,5)    # Let's check the shape of the Cauchy distribution
```

Unlike R, in Stan code, it is more common to specify the standard deviation rather than the variance.

model4.stan

```
1   data{
2     array[100] real y;  // Data
3   }
4   parameters{
5     real mu;  // Mean value
6     real<lower=0> sigma;  // Standard deviation
7   }
8   model{
9     mu ~ normal(0,100);  // Prior distribution of the mean value
10    sigma ~ cauchy(0,5);  // Prior distribution of the standard deviation
11    for(n in 1:100){
12      y[n] ~ normal(mu, sigma);  // Repeat 100 times with for
13    }
14  }
```

```
1  set.seed(456)
2  y <- rnorm(100,4,2)
3  hist(y)
4  model4 <- stan_model("model4.stan")
5  data_list <- list(y=y)
6  fit4 <- sampling(model4, data=data_list)
7  fit4
```

We can also make the number of elements in the array N an input variable:

model5.stan

```
1   data{
2     int N;  // Number of data
3     array[N] real y;  // Data
4   }
5   // The parameters block is the same
6   model{
7     mu ~ normal(0,100);  // Prior distribution of the mean value
8     sigma ~ cauchy(0,5);  // Prior distribution of the standard deviation
9     for(n in 1:N)
10      y[n] ~ normal(mu, sigma);  // Repeat N times with for
11  // In the case of a for loop with only one line of processing, {} can be
        omitted
12  }
```

```
1   model5 <- stan_model("model5.stan")
2   data_list <- list(N=100, y=y)
3   fit5 <- sampling(model5, data=data_list)
4   fit5
```

For loops can also be vectorized:

model6.stan

```
1   // data, parameters blocks are the same
2   model{
3     mu ~ normal(0,100);  // Prior distribution of the mean value
4     sigma ~ cauchy(0,5);  // Prior distribution of the standard deviation
5     y ~ normal(mu, sigma);  // Vectorization
6   }
```

This is a process that makes the N-dimensional y follow the normal() distribution at once. By doing so, some of the processes for calculating the likelihood of the normal distribution can be shared, making the calculation faster.

As we have seen so far, when using a new variable, it is necessary to specify the type of the variable. Continuous values are real type (real), and discrete values are integer type (int). However, note that the corresponding type varies depending on the distribution. For example, variables following continuous distributions such as normal distribution and Beta distribution need to use real, while variables following discrete distributions such as binomial distribution need to use int.

Vectors (vector) and matrices (matrix) can be specified to apply matrix operations (linear algebra operations). Also, parameters must be of type real, vector, or matrix.

For example, in multivariate normal distribution, data and mean use `vector`, and covariance matrix uses `matrix`. When specifying them concretely, write in the order of range, number of elements, and variable, as follows:

```
vector[2] x;
matrix[4,5] x;
vector<lower=0>[5] sigma;
```

In addition, there are arrays that are similar but different from these:

```
array[5] real x;
// A 1-dimensional array with 5 real numbers
array[5,5] int x;
// A 2-dimensional array with 5 X 5 integers
array[5] vector[2] x;
// An array containing 5 vectors, each with 2 elements
```

Arrays do not allow matrix operations, but for dependent variables' data assuming discrete distribution, it is necessary to use arrays of type `int`.

3.3.3 Simple Linear Regression

Simple linear regression is a problem of estimating the intercept and slope $\alpha, \beta \in \mathbb{R}$, and the variance of the noise $\sigma^2 > 0$, assuming that

$$y_i = \alpha + \beta x_i + e_i \ , \ e_i \sim N(0, \sigma^2)$$

holds independently for each $i = 1, \ldots, n$ from observations $x_1, \ldots, x_n \in \mathbb{R}, y_1, \ldots, y_n \in \mathbb{R}$. It is possible to calculate $\alpha, \beta \in \mathbb{R}, \sigma^2 \neq 0$ from the least squares method, but this corresponds to the case when estimated by the maximum likelihood method. Here, we use Stan to find the posterior distribution of $\alpha, \beta \in \mathbb{R}, \sigma > 0$ given $x_1, \ldots, x_n, y_1, \ldots, y_n \in \mathbb{R}$.

In the following, we model that α, β, σ^2 are generated based on the prior distribution, and for each $i = 1, \ldots, n, y_i \sim N(\alpha + x_i\beta, \sigma^2)$ holds:

model7.stan

```
 1   data{
 2     int N; // Sample size
 3     vector[N] y; // Dependent variable
 4     vector[N] x; // Explanatory variable
 5   }
 6     parameters{
 7     real alpha; // Intercept
 8     real beta; // Slope
 9     real <lower=0> sigma; // Standard deviation of the residuals
10   }
11     model{
```

Table 3.1 The response medv and 13 covariates in the Boston data

Column	Variable	Meaning of the variable
1	crim	Per capita crime rate by town
2	zn	Proportion of residential land zoned for lots over 25,000 sq.ft.
3	indus	Proportion of non-retail business acres per town
4	chas	Charles River dummy variable (= 1 if tract bounds river; 0 otherwise)
5	nox	Nox concentration (parts per 10 million)
6	rm	Average number of rooms per dwelling
7	age	Proportion of owner-occupied units built prior to 1940
8	dis	Weighted distances to five Boston employment centers
9	rad	Index of accessibility to radial highways
10	tax	Fixed asset tax rate per 10,000 dollars
11	ptratio	Pupil-teacher ratio by town
12	b	Proportion of blacks by town
13	lstat	Percentage of lower status of the population
14	medv	Median housing price in units of 1,000 dollars

```
12   alpha ~ normal(0,100);
13   beta ~ normal(0,100);
14   sigma ~ cauchy(0,5);
15   y ~ normal(alpha + beta * x, sigma); // x is declared as a vector
16   }
```

We examined the linear regression (simple regression) between the average number of rooms per household rm and the average housing price medv in the Boston dataset (Table 3.1) available in the CRAN MASS package. We built a model medv ~ normal(alpha + beta*rm, sigma) and obtained the posterior distributions of the intercept alpha, slope beta, and noise (square root of the residuals) sigma:

```
1   library(rstan)
2   library(MASS)
3   data_list <- list(N = nrow(Boston), y = Boston$medv, x = Boston$rm)
4   fit7 <- stan("model7.stan", data = data_list)
5   print(fit7, probs = c(0.025, 0.5, 0.975))
```

....

```
> print(fit7, probs = c(0.025, 0.5, 0.975))
Inference for Stan model: anon_model.
4 chains, each with iter=2000; warmup=1000; thin=1;
post-warmup draws per chain=1000, total post-warmup draws=4000.

         mean se_mean   sd     2.5%      50%    97.5% n_eff Rhat
alpha  -34.43    0.08 2.70   -39.63   -34.48   -29.20  1089    1
beta     9.06    0.01 0.43     8.23     9.07     9.89  1084    1
sigma    6.64    0.01 0.21     6.24     6.63     7.08  1650    1
lp__ -1208.83    0.04 1.27 -1212.12 -1208.51 -1207.40   902    1
```

If you want to output only specific parameters, do as follows:

```
print(fit7, pars=c("alpha", "beta"))
print(fit7, pars="sigma", digit=3)  # Specify the number of output digits
```

```
> print(fit7, pars=c("alpha", "beta"))

.........

          mean se_mean   sd    2.5%    25%    50%    75%  97.5% n_eff Rhat
alpha -34.43    0.08 2.70 -39.63 -36.29 -34.48 -32.57 -29.20  1089    1
beta    9.06    0.01 0.43   8.23   8.77   9.07   9.36   9.89  1084    1

.........

> print(fit7, pars="sigma", digit=3)

.........

         mean se_mean    sd  2.5%   25%   50%   75% 97.5% n_eff  Rhat
sigma 6.636   0.005 0.213 6.241 6.492 6.627 6.772 7.081  1650 1.003
```

Next, we consider finding the distribution of predicted values using the posterior probabilities of the generated parameters. We want to find the distribution of the value of y at a point $x_* \in \mathbb{R}$ different from the $x_1, \ldots, x_n \in \mathbb{R}$ used to estimate α, β. Without using Bayes, we find the confidence intervals for α, β and then find the confidence interval for $\alpha + x_*\beta$. Also, for each $\alpha, \beta \in \mathbb{R}$, $\alpha + x_*\beta$ can be calculated, but in reality, it fluctuates more than the confidence interval due to the independently varying variance σ^2 noise. It is also possible to calculate the confidence interval (prediction interval) that takes this effect into account.[3] In Bayes, the distribution of $\alpha + x_*\beta$ corresponding to the confidence interval is determined by the posterior distribution of α, β. Furthermore, another distribution corresponding to the prediction interval, considering the independent noise effect, is determined. Below, the `normal_rng()` function generates random numbers following a normal distribution:

model8.stan

```
data{
  int N; // Sample size
  vector[N] y; // Dependent variable
  vector[N] x; // Explanatory variable
  int N_pred; // The number of x to predict
  vector[N_pred] x_pred; // The vector consisting of the values of x to
      predict
}
// The data, parameters blocks are the same
```

---

[3] The details are described in Chap. 2 of *Statistical Machine Learning with Math and R* (Springer) by Joe Suzuki [14].

```
 9  generated quantities{
10    vector[N_pred] y_mu;
11    vector[N_pred] y_pred;
12    y_mu = alpha + beta * x_pred;      // Predicted values
13    for(n in 1:N_pred)
14      y_pred[n] = normal_rng(y_mu[n], sigma);  // Values with added noise
15      to the predicted values
16  }
```

We tried to predict the average housing price when the average number of rooms per household was 2, 3, 4, 5, 6:

```
1  rm_pred <- 2:6
2  data_list <- list(N = nrow(Boston), y = Boston$medv,
3    x = Boston$rm, N_pred = length(rm_pred), x_pred = rm_pred)
4  fit8 <- stan("model8.stan", data = data_list)
5  stan_dens(fit8, pars=c("y_mu","y_pred"))
```

We obtained the output as in Fig. 3.8.

Below, we introduce the transformed parameters block. This makes the processing easier to see and easier to extend:

model19.stan

```
1  // The data and parameters blocks are the same as in model7.stan,
2  transformed parameters{
3    vector [N] mu;
4    mu = alpha + beta * x; // x is declared as a vector
5  }
6  model{
7    alpha ~ normal(0,100);
8    beta ~ normal(0,100);
9    sigma ~ cauchy(0,5);
10   y ~ normal(mu, sigma);
11 }
```

In fact, by adding a generated quantities block, the likelihood of the parameters

$$-\log f(y_i|\mu_j, \sigma_j) , \ i = 1, \ldots, n , \ j = 1, \ldots, m$$

can be calculated with $f(\cdot|\mu, \sigma)$, being the probability density function of $N(\mu, \sigma^2)$. μ, σ follow the posterior distribution. For each sampling $j = 1, \ldots, m$, a different $\theta_j = (\mu_j, \sigma_j)$ is obtained. Since these are the realized values of $\theta = (\mu, \sigma)$ generated according to the posterior distribution $p(\theta|y_1, \ldots, y_n)$, we obtain

$$\int_\Theta f(y_i|\theta) p(\theta|y_1, \ldots, y_n) d\theta \approx \frac{1}{m} \sum_{j=1}^{m} f(y_i|\mu_j, \sigma_j). \tag{3.6}$$

model19.stan

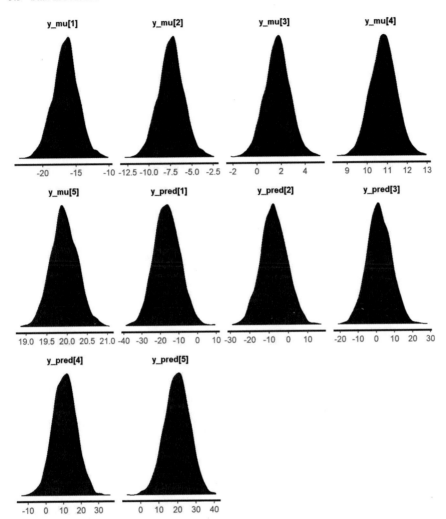

Fig. 3.8 The output of the mean housing price predictions

```
1  // Add the following generated quantities block
2  generated quantities{
3    array[N] real log_lik;
4    for(n in 1:N)
5      log_lik[n]= normal_lpdf(y[n]|mu,sigma);
6  }
```

Here, even if f is not a normal distribution, by appending _lpdf at the end, such as cauchy_lpdf or beta_lpdf, it means the logarithm of the probability density function. It is common to place parameters after $y_{[n]}|$. Also, _lpdf cannot be vectorized because its arguments are a mix of vectors and scalars. For example,

```
generated quantities{
array[N] real log_lik;
log_lik= normal_lpdf(y|mu,sigma)
}
```

will result in an error. Next, let's output log_lik:

```
1  y <- rnorm(100,5,2)
2  x <- rnorm(100,7,6)
3  model9 <- stan_model("model9.stan")
4  fit9 <- sampling(model9, data=list(N=100, x=x, y=y))
5  log_lik <- rstan::extract(fit9)$log_lik
6  is.matrix(log_lik)        # It is clear that it is a matrix
7  dim(log_lik)              # Number of samples X number of samplings
8  head(log_lik,5)           # The first 5 rows
```

....

```
> log_lik <- rstan::extract(fit9)$log_lik
> is.matrix(log_lik)
[1] TRUE
> dim(log_lik)
[1] 4000  100
> head(log_lik,5)

iterations     [,1]      [,2]      [,3]      [,4]      [,5]      [,6]      [,7]
      [1,] -3.403301 -1.657132 -1.725351 -2.779104 -2.936850 -2.715903 -1.917382
      [2,] -3.653273 -1.698815 -1.692000 -2.890423 -2.650699 -2.649536 -1.990583
      [3,] -3.622558 -1.629874 -1.682376 -2.834984 -2.848217 -2.820437 -1.908749
      [4,] -3.841320 -1.640801 -1.659828 -2.841350 -2.617469 -2.938005 -1.914694
      [5,] -4.102244 -1.746632 -1.650777 -2.999545 -2.293481 -2.712034 -2.053173
```

....

```
iterations    [,99]     [,100]
      [1,] -1.747460 -1.926040
      [2,] -1.717556 -1.994222
      [3,] -1.715444 -1.913033
      [4,] -1.707978 -1.910116
      [5,] -1.684236 -2.042841
```

The right-hand side of (3.6) can be written in R language using the functions exp() and colMeans() as

```
colMeans(exp(log_lik))
```

Furthermore, taking the product for $i = 1, \ldots, n$ and then taking the logarithm

```
-mean(log(colMeans(exp(log_lik))))
```

corresponds to the empirical loss required to calculate WAIC.

3.3.4 Multiple Regression

Multiple regression is a problem of estimating the intercept and slope $\beta_0 \in \mathbb{R}$, $\beta \in \mathbb{R}^p$, and the noise variance $\sigma^2 > 0$ under the assumption that for each $i = 1, \ldots, n$, the following independent relationship holds:

$$y_i = \beta_0 + x_i \beta + e_i \,, \ e_i \sim N(0, \sigma^2 I_n)$$

given observations $x_1, \ldots, x_n \in \mathbb{R}^p$, $y_1, \ldots, y_n \in \mathbb{R}$, where $x_i \beta$ represents $\sum_{j=1}^{p} x_{i,j} \beta_j$ with $\beta = [\beta_1, \ldots, \beta_p]^\top$ (column vector) and $x_i = [x_{i,1}, \ldots, x_{i,p}]$ (row vector), and I_n is the identity matrix of size n.

In this case, let $X \in \mathbb{R}^{n \times (p+1)}$ be a matrix with a size n column vector with all components being 1 to the left of the matrix containing $x_1, \ldots, x_n$ in each row, and $y = [y_1, \ldots, y_n]^\top \in \mathbb{R}^n$. We denote $[\beta_0, \beta_1, \ldots, \beta_p]^\top$ as $\beta \in \mathbb{R}^{p+1}$ and use Stan to find the posterior distribution of $\beta \in \mathbb{R}^{p+1}$, $\sigma^2 > 0$ under X, y. In the following, we modeled that β, σ^2 are generated based on the prior distribution, and for each $i = 1, \ldots, n$, $y \sim N(X\beta, \sigma^2 I_n)$ holds for

$$y = \begin{bmatrix} y_1 \\ \vdots \\ y_n \end{bmatrix}, \ X = \begin{bmatrix} 1 & x_{1,1} & \cdots & x_{1,p} \\ \vdots & \vdots & \ddots & \vdots \\ 1 & x_{n,1} & \cdots & x_{n,p} \end{bmatrix}, \ \beta = \begin{bmatrix} \beta_0 \\ \beta_1 \\ \vdots \\ \beta_p \end{bmatrix}, \ \sigma^2 I_n = \begin{bmatrix} \sigma^2 & \cdots & 0 \\ \vdots & \ddots & \vdots \\ 0 & \cdots & \sigma^2 \end{bmatrix}.$$

model10.stan

```
data{
    int N; // Sample size
    int M; // Number of variables (including the intercept)
    vector[N] y; // Dependent variable
    matrix[N,M] x; // Independent variables, declared as matrix
}
parameters{
    vector[M] beta; // declared as vector
    real <lower=0> sigma;
}
model{
    beta ~ normal(0,100);
    sigma ~ cauchy(0,5);
    y ~ normal(x * beta, sigma);
}
```

```
x1 <- rnorm(100,0,1)
x2 <- rnorm(100,0,1)
y <- 2 + 5 * x1 + 7 * x2 +rnorm(100,0,3)
summary(lm(y~x1+x2))
intercept <- rep(1,100)
x <- data.frame(intercept, x1, x2)
head(x)
```

```
 8  data_list <- list(N=100, M=3, y=y, x=x)   # Note M=3
 9  model10 <- stan_model("model10.stan")
10  fit10 <- sampling(model10, data=data_list)
11  print(fit10, probs=c(0.025, 0.5, 0.975))
```

In particular, summary, head, and the last print output as follows:

```
> summary(lm(y~x1+x2))

Call:
lm(formula = y ~ x1 + x2)

Residuals:
    Min      1Q  Median      3Q     Max
-7.8477 -1.9461  0.0227  1.8287  6.6660

Coefficients:
            Estimate Std. Error t value Pr(>|t|)
(Intercept)   2.1942     0.2866   7.657 1.43e-11 ***
x1            4.9308     0.2636  18.706  < 2e-16 ***
x2            6.7242     0.2869  23.437  < 2e-16 ***
---
Signif. codes:  0 '***' 0.001 '**' 0.01 '*' 0.05 '.' 0.1 ' ' 1

Residual standard error: 2.836 on 97 degrees of freedom
Multiple R-squared:  0.9036,    Adjusted R-squared:  0.9016
F-statistic: 454.5 on 2 and 97 DF,  p-value: < 2.2e-16


.........


> head(x)
  intercept         x1         x2
1         1 -0.9823585  0.7517814
2         1  1.7503470 -1.3702254
3         1  0.9805008  0.1549586
4         1 -1.0044848  0.5919998
5         1  1.8609173 -0.9818162
6         1 -0.3310599 -0.2195880


.........


> print(fit10, probs=c(0.025, 0.5, 0.975))
Inference for Stan model: anon_model.
4 chains, each with iter=2000; warmup=1000; thin=1;
post-warmup draws per chain=1000, total post-warmup draws=4000.

          mean se_mean   sd    2.5%     50%   97.5% n_eff Rhat
beta[1]   2.19    0.00 0.30    1.61    2.20    2.77  4285    1
beta[2]   4.93    0.00 0.27    4.41    4.93    5.44  4150    1
beta[3]   6.72    0.00 0.29    6.15    6.72    7.30  3789    1
sigma     2.87    0.00 0.21    2.50    2.86    3.31  3637    1
lp__   -154.02    0.03 1.45 -157.74 -153.70 -152.18  1910    1
```

We can also calculate the empirical loss by adding generated quantities to model10.stan:

model11.stan

```
1  // Add the following generated quantities block below
2  generated quantities{
3    array[N] real log_lik;
4    for(n in 1:N)
5      log_lik[n]= normal_lpdf(y[n]|x[n]*beta, sigma);
6  }
```

3.3.5 Mixture of Normal Distributions

For example, as shown when executing the last line above, in addition to the specified parameters, the parameter `lp_` (two underscores after `lp`) is automatically generated. This value corresponds to the posterior log-likelihood, which, for $\theta = (\beta, \sigma^2)$, is the logarithm of the likelihood $\varphi(\theta) \prod_{i=1}^{n} p(x_i|\theta)$:

$$\log \varphi(\theta) + \sum_{i=1}^{n} \log p(x_i|\theta).$$

The posterior log-likelihood can also be used in the `model` block. In that case, use the variable name `target`.

Note that the following Stan code behaves the same as `model7.stan`:

model12.stan

```
1  // The data and parameters blocks are the same as in model7.stan
2  model{
3    for(n in 1:N)
4      target += normal_lpdf(y[n] | alpha + beta * x[n], sigma);
5    target += normal_lpdf(alpha|0,100);
6    target += normal_lpdf(beta |0,100);
7    target += cauchy_lpdf(sigma|0,  5);
8  }
```

Here, `normal_lpdf` and `cauchy_lpdf` are the log-likelihoods when assuming a normal distribution and a Cauchy distribution, respectively. `target` is initially initialized to 0, and `target +=` is a shorthand for `target = target +`.

Using `target`, we can write the Stan code for a mixture of normal distributions. The probability density function of $y \in \mathbb{R}$ following a mixture of normal distributions is given by

$$\theta f(y|\mu_1, \sigma_1) + (1 - \theta) f(y|\mu_2, \sigma_2)$$

for $0 < \theta < 1, \mu_1 < \mu_2$, and $\sigma_1, \sigma_2 > 0$, where $f(y|\mu, \sigma)$ represents the probability density function of $y \in N(\mu, \sigma^2)$. The following Stan code allows the parameters μ_1, μ_2 to vary in the range $\mu_1 < \mu_2$. Also, it is common to use the function `log_mix()` provided by Stan when the number of classes is 2:

model13.stan

```
1   data {
2     int<lower = 0> N;
3     vector[N] y;
4   }
5   parameters {
6     ordered[2] mu;
7     real<lower=0, upper=1> theta;
8     real<lower=0> sigma;
9   }
10  model {
11    mu ~ normal(0, 2);
12    theta ~ beta(5, 5);
13    for (n in 1:N)
14      target += log_mix(theta,
15        normal_lpdf(y[n] | mu[1], sigma),
16        normal_lpdf(y[n] | mu[2], sigma));
17  }
```

It is fine when the number of classes is 2 (Rhat is slightly above 1), but it becomes unstable as the number of classes increases. The following execution performs parallel processing and uses the cores to their maximum capacity:

```
1   library(rstan)
2   rstan_options(auto_write = TRUE)
3   options(mc.cores = parallel::detectCores())  # package rstudioapi is
4             required
5   N <- 100
6   y <- rnorm(100)
7   data_list <- list(N = N, y = y)
8   fit13 <- stan(file = "model13.stan", data = data_list, seed = 1)
9     # seed = 1 will generate the same random numbers
10  stan_dens(fit13)
```

Appendix: Proof of Proposition

Proof of Proposition 3

At a certain time, when θ is given, the probability that θ' is generated and accepted is $s(\theta'|\theta) \min\{1, \exp(-H(\theta') + H(\theta))\}$. Integrating this with respect to θ' gives the conditional probability of acceptance for θ,

$$Q(\theta) := \int_\Theta s(\theta'|\theta) \min\left(1, \exp\{-H(\theta') + H(\theta)\}\right) d\theta'.$$

The probability of staying in the same state is $1 - Q(\theta)$, so

$$P(\theta'|\theta) = s(\theta'|\theta) \min\left\{1, \exp(-H(\theta') + H(\theta))\right\} + \delta(\theta' - \theta)\{1 - Q(\theta)\}, \quad (3.7)$$

where $\delta(x)$ is the delta function and $\int_\Theta \delta(\theta' - \theta)(1 - Q(\theta'))d\theta' = 1 - Q(\theta)$ holds.
From this and (3.3), we have

$$P(\theta'|\theta) \exp\{-H(\theta)\}$$
$$= s(\theta'|\theta) \min \{\exp(-H(\theta)), \exp(-H(\theta'))\} + \delta(\theta' - \theta)\{1 - Q(\theta)\} \exp\{-H(\theta)\}$$
$$= s(\theta|\theta') \min \{\exp(-H(\theta)), \exp(-H(\theta'))\} + \delta(\theta' - \theta)\{1 - Q(\theta')\} \exp\{-H(\theta')\}$$

$$(3.8)$$

$$= P(\theta|\theta') \exp\{-H(\theta')\} \,,$$

where we used the fact that the term $\delta(\theta' - \theta)\{1 - Q(\theta')\}$ in each equation is nonzero
only when $\theta = \theta'$. ∎

Exercises 14–26

14. Why does (3.7) hold in the proof of Proposition 3? What about equality (3.8)?
15. The function `lik` takes the parameters $\theta = (\theta_1, \theta_2) \in \Theta$ of a normal distribution
    and the samples $x_1, \ldots, x_n \in \mathcal{X}$, and outputs

$$\frac{1}{\sqrt{2\pi\theta_2}} \exp\{-\frac{(x_i - \theta_1)^2}{2\theta_2}\} \,, \ i = 1, \ldots, n.$$

What does it output? Also, what does the function `accept(x,y)` output for
the cases $y \geq x$ and $y < x$?
16. In the function `MCMC`, what values are stored in the variables `output` and
    `accept_reject` after execution? Also, in the execution

```
1   result = MCMC(metropolis_hastings, c(mu_obs,3), 50000, obs)
2   m <- 50000; # m=100
3   output <- result$output
4   output2 <- output[seq(2,2*m,2)]
```

what values are stored in the variables `output` and `output2`?
17. Prove the conservation of energy from (3.4) and (3.5).
18. In Example 22, when $\theta(t) = \sin t$ and $p(t) = \cos t$, how should the functions
    `euler` and `leapfrog` be changed? Also, execute and output a graph like Fig. 3.5.
19. Generate $x_1, \ldots, x_n \in \mathbb{R}$ ($n = 100$) according to the standard normal distribu-
    tion, and confirm in Stan that under the conditions of Example 4, the posterior
    probability of μ becomes (2.11) (compare the theoretical posterior distribution
    written in R language with the curve output by Stan's `stan_dens()`).
20. Find the posterior distribution for the case of a Beta distribution with parameters
    (a, b) as a prior distribution and D occurrences out of N trials (using a formula,
    not Stan).

21. It is known that the least squares solution $\hat{\beta} = (X^\top X)^{-1} X^\top y \in \mathbb{R}^{p+1}$ follows $N(\beta, \sigma^2 (X^\top X)^{-1})$. Show that the variance of the inner product $x_* \hat{\beta}$ with row vector $x_* \in \mathbb{R}^{p+1}$, whose first component is 1, is $\sigma^2 x_* (X^\top X)^{-1} x_*^\top$, and that the variance of $x_* \hat{\beta} + e$, with $e \sim N(0, \sigma^2)$ added independently of $x_* \hat{\beta}$, is $\sigma^2 \{ x_* (X^\top X)^{-1} x_*^\top + 1 \}$.

22. Add the covariate LSTAT (proportion of low-income population) to the Boston data, use the model10.stan file to calculate the posterior distribution, and display the posterior distribution of all parameters using the function stan_hist.

23. Apply model10.stan instead of model7.stan to the following process and find the same solution:

```
1  library(rstan)
2  library(MASS)
3  data_list <- list(N = nrow(Boston), y = Boston$medv, x = Boston$rm)
4  fit7 <- stan("model7.stan", data = data_list)
5  print(fit7, probs = c(0.025, 0.5, 0.975))
```

24. We want to make the standard deviations different as well as the means of the two classes of the mixed normal distribution. Create Stan code 24.stan and execute the following:

```
1  library(rstan)
2  N <- 100
3  y <- rnorm(100)
4  data_list <- list(N = N, y = y)
5  fit <- stan(file = "24.stan", data = data_list)
6  stan_dens(fit)
```

25. We want to find the posterior distribution of the difference in means from two types of samples with the same number of samples. We want to find the posterior distribution of mu_x - mu_y. Complete the generated quantities, display the posterior distribution, and find the probability of $\mu_x > \mu_y$:

```
1  data{
2    int N; // Sample size
3    array[N] real y; // Data 1
4    array[N] real x; // Data 2
5  }
6  parameters{
7    real mu_y; // mean of y
8    real <lower=0> sigma_y; // standard deviation of y
9    real mu_x; // mean of x
10   real <lower=0> sigma_x; // standard deviation of x
11  }
12  model{
13    mu_y ~ normal(0,100);    // prior distribution of the mean of y
14    sigma_y ~ cauchy(0,5);   // prior distribution of the standard
          deviation of y
15    mu_x ~ normal(0,100);    // prior distribution of the mean of x
16    sigma_x ~ cauchy(0,5);   // prior distribution of the standard
          deviation of x
17    y ~ normal(mu_y,sigma_y);
```

```
18   // y follows a normal distribution with mean mu_y and standard
         deviation sigma_y
19   x ~ normal(mu_x,sigma_x);
20   // x follows a normal distribution with mean mu_x and standard
         deviation sigma_x
21   }
22   generated quantities{
23   // To be completed
24   }
```

(Assuming the filename above is model_X.stan)

```
1   y <- rnorm(100,5,2)
2   x <- rnorm(100,7,6)
3   fit_X <- stan("model_X.stan", data=list(N=100,y=y,x=x))
4   diff <- rstan:: extract(fit_X)$diff # Random numbers following the
         posterior distribution of the difference
5   plot(density(diff))                # Display the posterior
6        distribution of the difference
```

26. The following is Stan code to find the posterior distribution of the parameters for logistic regression:

```
1   data{
2     int N; // Sample size
3     int M; // Number of variables (including intercept)
4     array[N] int y; // Target variable
5     matrix[N,M] x; // Explanatory variables, declared as matrix
6   }
7   parameters{
8     vector[M] beta; // Declared as vector
9   }
10  model{
11    beta ~ normal(0,100);
12    y ~ bernoulli_logit(x*beta);  // Shorter code and faster
         estimation
13  }
```

Actually, with $N = 100$ and $M = 3$, generate random data for the covariates and the response, and display the posterior distribution of each parameter using stan_dens.

Chapter 4
Mathematical Preparation

In this chapter, we describe the mathematical knowledge necessary for understanding this book. First, we discuss matrices, open sets, closed sets, compact sets, the Mean Value Theorem, and Taylor expansions. All of these are topics covered in the first year of college. Next, we discuss absolute convergence and analytic functions. Then, we discuss the Law of Large Numbers and the Central Limit Theorem, as well as defining the symbols $O_P(\cdot)$ and $o_P(\cdot)$ used in subsequent chapters. Finally, we define the Fisher information matrix and discuss the properties of regular and realizable cases. For algebraic geometry and related topics, please refer to Chap. 6. Readers who already understand the content of this chapter may skip it as appropriate. At the end of the chapter, we provide the proof of Proposition 2, which was postponed in Chap. 1. It is assumed that with the preliminary knowledge of this chapter, it can be understood.

4.1 Elementary Mathematics

Here we discuss matrices and eigenvalues, open sets, closed sets, compact sets, the Mean Value Theorem, and Taylor expansions.

4.1.1 Matrices and Eigenvalues

A matrix $A \in \mathbb{R}^{n \times n}$ ($n \geq 1$) with the same number of rows and columns is called a *square matrix*. A *diagonal matrix* is a matrix whose off-diagonal elements are all zero. A diagonal matrix with all diagonal elements equal to 1 is called an *identity matrix*, denoted as $I_n \in \mathbb{R}^{n \times n}$. The sum of the diagonal elements of a square matrix is

© The Author(s), under exclusive license to Springer Nature Singapore Pte Ltd. 2023
J. Suzuki, *WAIC and WBIC with R Stan*,
https://doi.org/10.1007/978-981-99-3838-4_4

called the *trace*. A matrix with zero elements in the (i, j) $(i < j)$ positions is called a *lower triangular matrix*.

In the following, for a square matrix $A \in \mathbb{R}^{n \times n}$, we assume that there exists an $X \in \mathbb{R}^{n \times n}$ such that $AX = I_n$, and we try to find it. To do this, we perform two types of operations on the matrix $[A \mid I_n] \in \mathbb{R}^{n \times 2n}$, which consists of A and I_n arranged side by side:

1. Subtract a multiple of one row from another row
2. Swap two rows

We obtain a matrix such that the left half becomes a lower triangular matrix. Assuming that we performed operation 2 a total of m times, the product of the diagonal elements of the left half of the matrix at this point, multiplied by $(-1)^m$, is called the *determinant* of the matrix A. In the following, we write the determinant of the matrix A as det A.

After performing these operations, initially with $B = I_n$, $[A \mid B]$ is transformed into $[A' \mid B']$, but X satisfying $AX = B$ also satisfies $A'X = B'$. If the determinant of A is not zero, we perform the above two operations further to make the left half a diagonal matrix. Finally, by

1. Dividing each row by the value of the diagonal element

we make the left half the identity matrix I_n.[1] If $A'' = I_n$, then for $[A'' \mid B'']$, $A''X = B''$, so the right half B'' at that time is X. Conversely, if the determinant of A is zero, such a matrix X does not exist. When a square matrix exists such that $AX = I_n$ (when the determinant of A is not zero), X is called the *inverse matrix* of A, denoted as $X = A^{-1}$.

Example 24 In each of the cases $d \neq 0$ and $d = 0$,

$$
\begin{bmatrix} a & b \\ c & d \end{bmatrix} \rightarrow \begin{bmatrix} a - bc/d & 0 \\ c & d \end{bmatrix}, \quad \begin{bmatrix} a & b \\ c & 0 \end{bmatrix} \rightarrow \begin{bmatrix} c & 0 \\ a & b \end{bmatrix}
$$

can be done. The determinant is $(a - bc/d) \cdot d \cdot (-1)^0$ for $d \neq 0$, and $cb \cdot (-1)^1$ for $d = 0$, both of which can be seen to be $ad - bc$. ∎

Example 25 When the determinant of A, $\Delta = ad - bc$, is not zero, in particular when $d \neq 0$,

$$
\begin{bmatrix} a & b & 1 & 0 \\ c & d & 0 & 1 \end{bmatrix} \rightarrow \begin{bmatrix} a - bc/d & 0 & 1 & -b/d \\ c & d & 0 & 1 \end{bmatrix} \rightarrow \begin{bmatrix} \Delta/d & 0 & 1 & -b/d \\ 0 & d & -cd/\Delta & 1 + bc/\Delta \end{bmatrix}
$$
$$
\rightarrow \begin{bmatrix} 1 & 0 & d/\Delta & -b/\Delta \\ 0 & 1 & -c/\Delta & a/\Delta \end{bmatrix}
$$

can be done, and

---

[1] The method of obtaining the inverse matrix by operations 1, 2, and 3 is called Gaussian elimination.

$$\begin{bmatrix} a & b \\ c & d \end{bmatrix} \cdot \frac{1}{\Delta} \begin{bmatrix} d & -b \\ -c & a \end{bmatrix} = \begin{bmatrix} 1 & 0 \\ 0 & 1 \end{bmatrix}$$

holds. That is, $\frac{1}{\Delta} \begin{bmatrix} d & -b \\ -c & a \end{bmatrix}$ becomes the inverse matrix of $\begin{bmatrix} a & b \\ c & d \end{bmatrix}$. ∎

Moreover, when a constant $\lambda \in \mathbb{C}$ and a vector $u \in \mathbb{C}^n$ ($u \neq 0$) exist such that $Au = \lambda u$, λ is called an *eigenvalue*, and u is called an *eigenvector*. If the matrix $A - \lambda I_n$ has an inverse, that is, if the determinant of $A - \lambda I_n$ is not zero, then from $u = (A - \lambda I_n)^{-1} 0 = 0$, the u that satisfies $Au = \lambda u$ is limited to $u = 0$. Eigenvalues are determined as solutions to the equation concerning λ (eigenvalue equation) stating that the determinant of $A - \lambda I_n$ is zero. In other words,

$$Au = \lambda u, u \neq 0 \Longleftrightarrow (A - \lambda I_n)u = 0, u \neq 0 \Longleftrightarrow \det(A - \lambda I_n) = 0$$

holds.

Example 26 For $n = 2$, if we set $A = \begin{bmatrix} a & b \\ c & d \end{bmatrix}$, then $\det(A - \lambda I_2) = (a - \lambda)(d - \lambda) - bc = 0$ holds. Therefore, the solutions of the quadratic equation $\lambda^2 - (a + d)\lambda + ad - bc = 0$ are the eigenvalues. ∎

A matrix $A \in \mathbb{R}^{n \times n}$ for which all the (i, j) components $A_{i,j}$ and (j, i) components $A_{j,i}$ are equal is called a *symmetric matrix*. In general, eigenvalues λ are not necessarily real numbers, but when the matrix $A \in \mathbb{R}^{n \times n}$ is symmetric, λ becomes a real number. In fact, for $\lambda \in \mathbb{C}$ and $u \in \mathbb{C}^n$ ($u \neq 0$), since[2] $A\bar{u} = \overline{Au} = \overline{\lambda u} = \bar{\lambda}\bar{u}$, we have

$$\langle Au, \bar{u} \rangle = \langle \lambda u, \bar{u} \rangle = \lambda \langle u, \bar{u} \rangle$$

and

$$\langle Au, \bar{u} \rangle = \langle u, A\bar{u} \rangle = \langle u, \overline{Au} \rangle = \langle u, \overline{\lambda u} \rangle = \langle u, \bar{\lambda}\bar{u} \rangle = \bar{\lambda}\langle u, \bar{u} \rangle,$$

where $\bar{z}$ denotes the *complex conjugate* of $z \in \mathbb{C}$, and for $a, b \in \mathbb{R}$, we set $\overline{a + ib} = a - ib$.

Example 27 For the matrix $\begin{bmatrix} a & b \\ c & d \end{bmatrix}$, if we set $b = c$, the eigenvalue equation becomes $\lambda^2 - (a + d)\lambda + ad - b^2 = 0$, and its discriminant is $(a + d)^2 - 4(ad - b^2) = (a - d)^2 + 4b^2 \geq 0$. Indeed, the eigenvalues are real numbers. ∎

For a symmetric matrix A, it is called *non-negative definite* when all eigenvalues are non-negative, and *positive definite* when all eigenvalues are positive.

---

[2] For $u, v \in \mathbb{C}, \overline{uv} = \bar{u} \cdot \bar{v}$ holds.

Example 28 For the matrix $\begin{bmatrix} a & b \\ c & d \end{bmatrix}$, if we set $b = c$, when $a + d \geq 0$, and $ad \geq b^2$, the two solutions of the eigenvalue equation are non-negative, and it becomes non-negative definite. Furthermore, if both eigenvalues are positive, that is, $a + d \geq 0$, and $ad > b^2$, it becomes positive definite.   ∎

Moreover, for a symmetric matrix $A \in \mathbb{R}^{n \times n}$, $z^\top A z, z \in \mathbb{R}^n$ is called the quadratic form of A.

Proposition 4 *A symmetric matrix $A \in \mathbb{R}^{n \times n}$ being non-negative definite is equivalent to the quadratic form $z^\top A z$ being non-negative for any $z \in \mathbb{R}^n$. Furthermore, A being positive definite is equivalent to the quadratic form $z^\top A z$ being positive for any $0 \neq z \in \mathbb{R}^n$.*

For the proof, please refer to the appendix at the end of the chapter.

4.1.2 Open Sets, Closed Sets, and Compact Sets

Let the Euclidean distance between each $x, y \in \mathbb{R}^d$ be denoted as $dist(x, y)$. For a subset M of $\mathbb{R}^d$, let us denote the open ball (excluding the boundary) of radius $\epsilon > 0$ centered at $z \in M$ as $B(z, \epsilon) := \{y \in \mathbb{R}^d \mid dist(z, y) < \epsilon\}$. If there exists a radius $\epsilon > 0$ such that $B(z, \epsilon) \subseteq M$ for any $z \in M$, M is called an *open set*. On the other hand, for any $\epsilon > 0$, if the intersection of $B(z, \epsilon)$ and M is non-empty, $z \in \mathbb{R}^d$ is called a *tactile point* of $M \subseteq \mathbb{R}^d$. If M contains all its tactile points as its elements, M is called a *closed set* (see Fig. 4.1). Generally, the complement of a closed set is an open set, and the complement of an open set is a closed set.

In fact, if M is a closed set, its tactile points are not included in the complement M^C, so when the radius of the open ball for each $z \in M^C$ is chosen to be small, the open ball will not intersect M. Conversely, if M is an open set, when the radius of the open ball for each $z \in M$ is chosen to be small, the open ball will not intersect M^C. Therefore, the tactile points of M^C are not in M.

Example 29 Assume that $d = 1$ and $d = 3$ for items 1–4 and item 5, respectively.

1. The open interval (a, b) is an open set, and the closed interval $[a, b]$ is a closed set.
2. The set of all real numbers $\mathbb{R}$ and the set of all integers $\mathbb{Z}$ are closed sets ($\mathbb{R}$ is also an open set).
3. The set $\mathbb{R} \cap \mathbb{Z}^C$, which is the set of all real numbers $\mathbb{R}$ excluding the set of all integers $\mathbb{Z}$, is an open set.
4. The set of all rational numbers $\mathbb{Q}$ is neither an open set nor a closed set.
5. The region $\{(x, y, z) \in \mathbb{R}^3 \mid x^2 + y^2 + z^2 < 1, z \geq 0\}$ is neither an open set nor a closed set.

∎

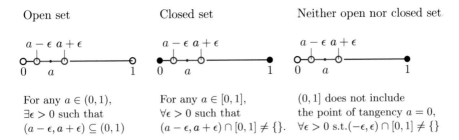

<table>
<tr><td>Open set</td><td>Closed set</td><td>Neither open nor closed set</td></tr>
</table>

For any $a \in (0, 1)$,
$\exists \epsilon > 0$ such that
$(a - \epsilon, a + \epsilon) \subseteq (0, 1)$

For any $a \in [0, 1]$,
$\forall \epsilon > 0$ such that
$(a - \epsilon, a + \epsilon) \cap [0, 1] \neq \{\}$.

$(0, 1]$ does not include
the point of tangency $a = 0$,
$\forall \epsilon > 0$ s.t. $(-\epsilon, \epsilon) \cap [0, 1] \neq \{\}$

Fig. 4.1 Open sets, closed sets, and cases that are neither. An open set is one where any point in the set is included if the neighborhood is made small enough. A tactile point is one that intersects the set no matter how small the neighborhood is made. A closed set is one that contains all tactile points

In addition, there is a concept of compact sets related to closed sets. When a mapping $M \ni x \mapsto \epsilon(x) \in \mathbb{R} > 0$ is arbitrarily defined and a finite number of $z_1, \ldots, z_m$ are used such that the union of open balls $\cup_{i=1}^{m} B(z_i, \epsilon(z_i))$ contains M as a subset, M is called *compact*. In this book, we only deal with subsets of $\mathbb{R}^d$ as the universal set and Euclidean distance as the distance. In this case, it is known that compact sets are equivalent to closed sets with bounded domains (*bounded closed sets*), where we say a set M is *bounded* when there exists a positive constant $L > 0$ such that $dist(x, y) < L$ for any $x, y \in M$. Among the closed sets in Example 29, $[a, b]$ is compact, but $\mathbb{R}$ and $\mathbb{Z}$ are not.

Although the proof is omitted, if a set M is compact, a continuous function with domain M has maximum and minimum values.

Example 30 $M = (0, 1], [1, \infty)$ are not compact. The continuous function $f(x) = 1/x$ does not have a maximum value on $M = (0, 1]$ and does not have a minimum value on $M = [1, \infty)$. ∎

Also, let $dist(x, a)$ be the distance between $x, a \in M$ ($M = \mathbb{R}$, for example, $dist(x, a) = |x - a|$). For any $\epsilon > 0$, a function f with domain M is said to be *continuous* (continuous) at $x = a$ if there exists a $\delta = \delta(\epsilon, a)$ such that:

$$dist(x, a) < \delta \Longrightarrow |f(x) - f(a)| < \epsilon.$$

If the function is continuous for all $a \in M$, then f is continuous.

The continuity of functions can be defined not only for $f : \mathbb{R} \to \mathbb{R}$. From Chap. 4 onwards, we will examine the set of continuous functions $C(K)$ defined on a compact set K. The distance between elements ϕ and ϕ' in $C(K)$ is defined by the *sup-norm* (uniform norm):

$$dist(\phi, \phi') := \sup_{\theta \in K} |\phi(\theta) - \phi'(\theta)|. \tag{4.1}$$

Then, the continuity of a function $f : C(K) \to \mathbb{R}$ at $\phi = \phi_a \in C(K)$ is defined by the existence of $\delta = \delta(\epsilon, \phi_a)$ such that for any $\epsilon > 0$,

$$dist(\phi, \phi_a) < \delta \implies |f(\phi) - f(\phi_a)| < \epsilon.$$

4.1.3   Mean Value Theorem and Taylor Expansion

In the following chapters, we will discuss the Mean Value Theorem and Taylor expansion, which will be used several times. They are particularly necessary for mathematical analysis when the sample size n is large.

The *Mean Value Theorem* asserts that for a differentiable function $f : \mathbb{R} \to \mathbb{R}$, if $a < b$, then there exists a c such that $a < c < b$ satisfying

$$\frac{f(b) - f(a)}{b - a} = f'(c). \tag{4.2}$$

Example 31 For $f(x) = x^2 - 3x + 2$, $a = 2$, and $b = 4$, we have

$$\frac{(b^2 - 3b + 2) - (a^2 - 3a + 2)}{b - a} = a + b - 3 = 3 \,, \quad f'(c) = 2c - 3.$$

So, $c = 3$ satisfies the condition. ∎

Equation (4.2) can be written as $f(b) = f(a) + f'(c)(b - a)$, which is an extended to Taylor's theorem.

Namely, if f is continuous up to the $(n - 1)$-th derivative and is n times differentiable,

$$f(b) = f(a) + \frac{f'(a)}{1!}(b - a) + \frac{f''(a)}{2!}(b - a)^2 + \cdots + \frac{f^{(n-1)}(a)}{(n - 1)!}(b - a)^{n-1} + R_n \tag{4.3}$$

with

$$R_n = \frac{f^{(n)}(c)}{n!}(b - a)^n.$$

There exists an $a < c < b$. If $n = 1$, it becomes the Mean Value Theorem. Sometimes it is written as $\theta a + (1 - \theta)b$ instead of c, and there exists such a $0 < \theta < 1$.

Setting $b = x$ in (4.3), we get

$$f(x) = f(a) + \frac{f'(a)}{1!}(x - a) + \frac{f''(a)}{2!}(x - a)^2 + \cdots + \frac{f^{(n-1)}(a)}{(n - 1)!}(x - a)^{n-1} + R_n$$

which is called the *Taylor expansion* of the function f at $x = a$. Furthermore, setting $a = 0$, we get

$$f(x) = f(0) + \frac{f'(0)}{1!}x + \frac{f''(0)}{2!}x^2 + \cdots + \frac{f^{(n-1)}(0)}{(n-1)!}x^{n-1} + R_n$$

which is called the *Maclaurin expansion*.

Example 32 When e^x and $\log(1 + x)$ are Maclaurin-expanded, there exist $0 < \theta < 1$ for each of

$$e^x = 1 + x + \frac{x^2}{2} + \cdots + \frac{x^{n-1}}{(n-1)!} + \frac{x^n}{n!}e^{\theta x} \tag{4.4}$$

and

$$\log(1 + x) = x - \frac{x^2}{2} + \frac{x^3}{3} - \cdots + (-1)^{n-2}\frac{x^{n-1}}{n-1} + (-1)^{n-1}\frac{x^n}{n} \cdot \frac{1}{(1+\theta x)^n}. \tag{4.5}$$

∎

For the case of two variables, a function $f : \mathbb{R}^2 \to \mathbb{R}$ that is continuous up to the $(n-1)$th derivative and differentiable n times, the Taylor expansion at $(x, y) = (a, b)$ can be written as

$$f(x, y) = \sum_{k=0}^{n-1}\sum_{i=0}^{k}\frac{1}{(k-i)!i!}(x-a)^i(y-b)^{k-i}\frac{\partial^k f}{\partial x^i \partial y^{k-i}}(a, b)$$

$$+ \sum_{i=0}^{n}\frac{1}{(n-i)!i!}(x-a)^i(y-b)^{n-i}\frac{\partial^n f}{\partial x^i \partial y^{n-i}}(\theta a + (1-\theta)x, \theta b + (1-\theta)y).$$

In the case of $n = 2$ for d variables, the Taylor expansion around $x = (x_1, \ldots, x_d)^\top = (a_1, \ldots, a_d)^\top = a$ is, when f has a continuous first derivative and is twice differentiable, written as

$$f(x) = f(a) + \sum_{i=1}^{d}(x_i - a_i)\frac{\partial f}{\partial x_i}(a) + \frac{1}{2}\sum_{i=1}^{d}\sum_{j=1}^{d}(x_i - a_i)(x_j - a_j)\frac{\partial^2 f}{\partial x_i \partial x_j}(\theta a + (1-\theta)x)$$

$$= f(a) + (x - a)^\top\{\nabla f(a)\} + \frac{1}{2}(x - a)^\top\{\nabla^2 f(\theta a + (1-\theta)x)\}(x - a),$$

where $\nabla f : \mathbb{R}^d \to \mathbb{R}^d$ is a vector consisting of the d partial derivatives of f, $\frac{\partial f}{\partial x_i}$, and $\nabla^2 f : \mathbb{R}^d \to \mathbb{R}^{d \times d}$ is a matrix (the Hessian matrix) consisting of the second partial derivatives of f, $\frac{\partial^2 f}{\partial x_i \partial x_j}$.

4.2   Analytic Functions

In the following, we will denote the set of non-negative integers by $\mathbb{N}$. Firstly, for $r = (r_1, \ldots, r_d) \in \mathbb{N}^d$, $x = (x_1, \ldots, x_d)$, $b = (b_1, \ldots, b_d) \in \mathbb{R}^d$, $a_r = a_{r_1, \ldots, r_d} \in \mathbb{R}$, we define

$$a_r(x - b)^r := a_{r_1, \ldots, r_d}(x_1 - b_1)^{r_1} \ldots (x_d - b_d)^{r_d}.$$

A sum of such terms

$$f(x) := \sum_{r \in \mathbb{N}^d} a_r(x - b)^r = \sum_{r_1 \in \mathbb{N}} \cdots \sum_{r_d \in \mathbb{N}} a_{r_1, \ldots, r_d}(x_1 - b_1)^{r_1} \ldots (x_d - b_d)^{r_d} , \quad x \in \mathbb{R}^d$$

(4.6)

is called a *power series*. When there are a finite number of non-zero terms, we call $f(x)$ a *polynomial with real coefficients* in terms of $x_1, \ldots, x_d$, and we denote the set of such polynomials as $\mathbb{R}[x]$ or $\mathbb{R}[x_1, \ldots, x_d]$. Furthermore, when there exists an open set U ($b \in U \subseteq \mathbb{R}^d$) such that for any $x \in U$, $\sum_r |a_r||x - b|^r < \infty$, we say that $f(x)$ *converges absolutely*. In this case, the infinite series (4.6) is independent of the order of the sums $\sum_{r_1}, \ldots, \sum_{r_d}$ and is unique. We call such a function $f : U \to \mathbb{R}$ an (real) *analytic function*.

Example 33   For the infinite series $\sum_{n=0}^{\infty} a_n$ with $a_n = (-1)^n$, we can write it in two ways:

$$(1 - 1) + (1 - 1) + \cdots = 0 + 0 + \cdots$$

and

$$1 - (1 - 1) - (1 - 1) - \cdots = 1 - 0 - 0 - \ldots,$$

which is due to the fact that $\sum_{n=0}^{\infty} |a_n| = 1 + 1 + \cdots = \infty$. However, in the case of $a_n = (-\frac{1}{2})^n$, we have $\sum_{n=0}^{\infty} |a_n| = 1 + \frac{1}{2} + \cdots = 2$, and hence the series converges.

■

Let a_n be a sequence of real numbers and $c \in \mathbb{R}$. When the power series $\sum_{n=0}^{\infty} a_n(x - c)^n$ converges absolutely if $|x - c| < R$ and diverges if $|x - c| > R$, we call R the radius of convergence (we need to investigate the case where $x - c$ equals the radius of convergence). If $a_n = 0$ except for a finite number of terms, $R := \lim_{n \to \infty} \left| \frac{a_n}{a_{n+1}} \right|$ will be the radius of convergence. In fact, if the absolute ratio of adjacent terms

$$r := \lim_{n \to \infty} \left| \frac{a_{n+1}(x - c)^{n+1}}{a_n(x - c)^n} \right| = \lim_{n \to \infty} \left| \frac{a_{n+1}}{a_n} \right| \cdot |x - c|$$

is $0 \leq r < 1$, it converges, and if $1 < r \leq \infty$, it diverges.

Example 34 For

$$f(x) = \sum_{n=1}^{\infty} \frac{1}{n} x^n = x + \frac{1}{2} x^2 + \frac{1}{3} x^3 + \cdots$$

the absolute ratio of adjacent terms is

$$\lim_{n \to \infty} \frac{|x|^{n+1}}{|x|^n} \frac{1/(n+1)}{1/n} = |x| \lim_{n \to \infty} \frac{n}{n+1} = |x|$$

for sufficiently large n. Therefore, it converges absolutely if $|x| < 1$. When investigating the case of $|x| = 1$, it becomes

$$\sum_{n=1}^{\infty} \frac{1}{n} = 1 + \frac{1}{2} + \frac{1}{3} + \cdots > \lim_{n \to \infty} \int_1^n \frac{dx}{x} = \lim_{n \to \infty} \log n = \infty,$$

so it does not converge absolutely when $|x| = 1$. Therefore, we can set the open set of the domain of f to be $U = (-1, 1)$. ■

Since taking the absolute value of each term makes it non-negative, absolute convergence becomes a convergence that does not assume the order of summation. However, what problems would arise with convergence that assumes the order of summation (conditional convergence)?

Example 35 If the series $\sum_{n=1}^{\infty} (-1)^{n-1} \frac{1}{n}$ is summed in the order of

$$1 - \frac{1}{2} + \frac{1}{3} - \frac{1}{4} + \cdots = \sum_{k=1}^{\infty} \frac{(-1)^{k-1}}{k}, \tag{4.7}$$

it becomes $\log 2$. In fact, if we denote the right-hand side of

$$\sum_{k=1}^{2n} \frac{(-1)^{k-1}}{k} = \left(\sum_{k=1}^{2n} \frac{(-1)^{k-1}}{k} + 2 \sum_{k=1}^{n} \frac{1}{2k} \right) - 2 \sum_{k=1}^{n} \frac{1}{2k} = \sum_{k=1}^{2n} \frac{1}{k} - \sum_{k=1}^{n} \frac{1}{k} = \sum_{k=1}^{n} \frac{1}{n+k}$$

as S_n, the equations

$$S_n = \frac{1}{n} \sum_{k=1}^{n} \frac{1}{1 + k/n} \leq \int_0^1 \frac{dx}{1 + x} \leq \frac{1}{n} \sum_{k=0}^{n-1} \frac{1}{1 + k/n} = S_n + \frac{1}{2n}$$

and

$$\int_0^1 \frac{dx}{1+x} - \frac{1}{2n} \le S_n \le \int_0^1 \frac{dx}{1+x} = \log 2$$

hold. On the other hand, if we first add the terms for $n = 1, 2, 4$, then the ones for odd numbers greater than or equal to 3, even numbers not divisible by 4 and greater than or equal to 6, and finally multiples of 4 greater than or equal to 8, (4.7) can be calculated as

$$\sum_{n=1}^{\infty} (-1)^{n-1} \frac{1}{n} = 1 - \frac{1}{2} - \frac{1}{4} + \sum_{n=2}^{\infty} \{ \frac{1}{2n-1} - \frac{1}{2(2n-1)} - \frac{1}{4n} \} = \frac{1}{4} + \frac{1}{2} \sum_{n=2}^{\infty} \{ \frac{1}{2n-1} - \frac{1}{2n} \}$$

$$= \frac{1}{2} \sum_{n=1}^{\infty} \{ \frac{1}{2n-1} - \frac{1}{2n} \} = \frac{1}{2} \lim_{n \to \infty} S_n = \frac{1}{2} \log 2.$$

∎

Although the proof is omitted, it is known that any series that converges conditionally can be made to converge to any real number by changing the order of its sum (Riemann's rearrangement theorem).

Example 36 In $U := \{(x, y) \in \mathbb{R}^2\}$, the series

$$\sum_{m=0}^{\infty} \sum_{n=0}^{\infty} \frac{x^m y^n}{m! n!}$$

converges absolutely. In fact, the ratio of the absolute values of any two adjacent terms converges to 0. Therefore, we can rearrange the order of the terms, and we obtain

$$\sum_{m=0}^{\infty} \frac{x^m}{m!} \cdot \sum_{n=0}^{\infty} \frac{y^n}{n!} = e^x \cdot e^y,$$

so the function $f : U \to \mathbb{R}$, $f(x, y) = e^{x+y}$, is an analytic function.   ∎

Here, if a function is differentiable any number of times $r \ge 0$ and the r-times differentiated function is continuous, the function is said to be of *class C^r*. On the other hand, analytic functions are continuous and differentiable, and no matter how many times they are differentiated, the asymptotic ratio of adjacent terms remains the same, making them analytic functions. That is, they are of *class C^∞*. Moreover, the analytic functions that can be expanded into power series can be uniquely expanded into Taylor series.

Note, however, that a function being of class C^∞ does not necessarily mean that it is an analytic function.

Example 37 The function

$$f(x) = \begin{cases} \exp(-1/x), & x > 0 \\ 0, & x \le 0 \end{cases}$$

is of class C^∞ but not analytic. In fact, the Taylor expansion at $x = 0$ results in $a_r = 0$, $r \in \mathbb{R}^d$ (Exercise 31), which contradicts the uniqueness of the Taylor expansion. ∎

In Chap. 8, we will assume that the average likelihood ratio $K(\theta) = \mathbb{E}_X[\log \frac{p(X|\theta_*)}{p(X|\theta)}]$ and the prior distribution $\varphi(\theta)$ are analytic functions on $\theta \in \Theta$ and proceed with the discussion. In this case, for the power series with real numbers a_r

$$\sum_{r \in \mathbb{N}^d} a_r (x - b)^r,$$

we considered whether $\sum_{r \in \mathbb{N}^d} |a_r| |(x - b)^r|$ is finite. In this book, we further assume that the likelihood ratio $f(x, \theta) = \log \frac{p(x|\theta_*)}{p(x|\theta)}$ is also an analytic function for $\theta \in \Theta$. However, in the case of multiple variables, preparations for extension are necessary. In this case, we consider as $a_r : \mathcal{X} \to \mathbb{R}$ and use the norm of a_r.

The set V with the properties

$$f, g \in V \Longrightarrow f + g \in V$$

and

$$\alpha \in \mathbb{R}, \ f \in V \to \alpha f \in V.$$

is called a *linear space*. In a linear space, we call $\| \cdot \| : V \to \mathbb{R}$ that satisfies the following conditions for each element a *norm* of V: for $\alpha \in \mathbb{R}$, $f, g \in V$

$$\|\alpha f\| = |\alpha| \cdot \|f\|, \ \|f + g\| \le \|f\| + \|g\|, \ \|f\| \ge 0, \ \|f\| = 0 \Longrightarrow f = 0.$$

In this book, we denote the set of $f : \mathcal{X} \to \mathbb{R}$ for which

$$\|f\|_2 := \sqrt{\int_{\mathcal{X}} \{f(x)\}^2 q(x) dx}$$

is finite as $L^2(q)$, where the true distribution q is used.

Here, the absolute value $| \cdot |$ becomes the norm of the one-dimensional Euclidean space $\mathbb{R}$, but the norm $\| \cdot \|_2$ also becomes the norm of the linear space $L^2(q)$ (problem 38). We often call it an analytic function taking real values when $a_r \in \mathbb{R}$ and an analytic function taking values in $L^2(q)$ when $a_r \in L^2(q)$, but in this book, we simply call the former an analytic function. Also, when we write each norm as $\| \cdot \|$, the set of x for which $\sum_{r \in \mathbb{N}^d} \|a_r\| |(x - b)^r|$ is finite becomes the domain.

For example, if the log-likelihood ratio $f(x, \theta)$

$$f(x, \theta) = \sum_{r \in \mathbb{N}^d} a_r(x)(\theta - \theta^1)^r$$

is an analytic function, it means that there exists a convergence domain (the radius of convergence is non-zero) such that

$$\sum_{r \in \mathbb{N}^d} \|a_r\|_2 |(\theta - \theta^1)^r| < \infty.$$

4.3 Law of Large Numbers and Central Limit Theorem

4.3.1 Random Variables

By preparing a universal set Ω and a set of its events in advance, when

$$\{\omega \in \Omega \mid X(\omega) \in O\}$$

becomes an event for any open set O of $\mathbb{R}$, we say that $X : \Omega \ni \omega \mapsto X(\omega) \in \mathbb{R}$ is *measurable*. Also, X is called a *random variable* that takes values in[3] $\mathbb{R}$. However, the way to determine the probability needs to be defined separately.

Example 38 When $\Omega = \{1, 2, 3, 4, 5, 6\}$ and $X(\omega) = (-1)^\omega$, it is necessary that at least $\{1, 3, 5\}$ and $\{2, 4, 6\}$ are events. That is, among the empty set $\{\}$ and the universal set Ω and these two sets, even if union, intersection, and complement operations are performed, no other than these four sets are generated. Also, by calculating the set of $\omega \in \Omega$ such that $X(\omega) \in (0, 1)$, the set of $\omega \in \Omega$ such that $X(\omega) \in (-2, 1)$, etc., we can see that the subset of Ω where $X(\omega) \in O$ for any open set O does not exist other than those four. The random variable X only defines the events, and the probability needs to be specified according to the axioms. ∎

Random variables can be defined not only as $\Omega \to \mathbb{R}$. If $\eta : \Omega \to C(K)$ is measurable, where $C(K)$ is a continuous function defined on a compact set K, then η is said to be a random variable that takes values in $C(K)$. Rather than considering it as a random variable, it can be seen as a random function. In defining measurability, open sets are defined using distance by the uniform norm.

4.3.2 Order Notation

First, we shall define the limit of a sequence of real numbers, which appears frequently in this book.

---

[3] It can be understood as a random variable.

An infinitely long sequence of real numbers a_n is said to *converge* to α as $n \to \infty$, or $\lim_{n \to \infty} a_n = \alpha$, if for any $\epsilon > 0$, $|a_n - \alpha| < \epsilon$ holds except for a finite number of n.[4]

Also, for a function $g(n)$ of positive integer n such as $g(n) = 1, n, n^2$, if $|g(n)a_n| < \epsilon$ holds for any $\epsilon > 0$ except for a finite number of n, i.e., if $g(n)a_n$ converges to 0 as $n \to \infty$, we write $a_n = o(\frac{1}{g(n)})$. On the other hand, if there exists an $M > 0$ such that $|g(n)a_n| < M$ holds except for a finite number of n, i.e., if $g(n)a_n$ is bounded, we write $a_n = O(\frac{1}{g(n)})$. For example, if it is $O(1/n)$, it is also $o(1)$.

4.3.3 Law of Large Numbers

Next, we will examine whether the sequence of probabilities $\{P(A_n)\}$ for a sequence of events $\{A_n\}$ converges to 1. When the probability $P(|X_n - \alpha| < \epsilon)$ converges to 1 as $n \to \infty$ for any $\epsilon > 0$, the sequence of random variables $\{X_n\}$ is said to *stochastically converge* to α, and we write it as $X_n \xrightarrow{P} \alpha$.

The Weak Law of Large Numbers is one of the most important theorems regarding stochastic convergence. Before introducing it, we shall show an important inequality.

Proposition 5 (Chebyshev's Inequality) *For a random variable with mean μ and variance $\sigma^2 > 0$, for any constant $k > 0$, the inequality*

$$P(|X - \mu| \geq k) \leq \sigma^2 / k^2$$

holds.

Proof Define I so that $I(A) = 1$ when event A occurs and $I(A) = 0$ otherwise. Then, the following inequality holds.

$$\sigma^2 = \mathbb{E}_X[(X - \mu)^2] \geq \mathbb{E}_X[(X - \mu)^2 I(|X - \mu| \geq k)] \geq k^2 \cdot P(|X - \mu| \geq k).$$

∎

Here, consider $\{X_n\}_{n=1}^{\infty}$ and $\{\epsilon_n\}_{n=1}^{\infty}$ as sequences of random variables. When

$$\frac{X_n}{\epsilon_n} \xrightarrow{P} 0$$

holds as $n \to \infty$, we write $X_n = o_P(\epsilon_n)$. Especially, when $X_n \xrightarrow{P} 0$ holds, we write $X_n = o_P(1)$.

Moreover, when there exist an $M > 0$ except for a finite number of n (they can depend on δ) such that

---

[4] It is equivalent to the existence of some $N(\epsilon)$ for any ϵ such that $|a_n - \alpha| < \epsilon$ for $n \geq N(\epsilon)$.

$$P\left(|X_n| \leq M\,|\epsilon_n|\right) \geq 1 - \delta$$

for any $\delta > 0$, we write

$$X_n = O_P\left(\epsilon_n\right).$$

Especially, if $P\left(|X_n| \leq M\right) \geq 1 - \delta$, we write $X_n = O_P(1)$.

o_P and O_P have the following properties for sequences of random variables $\{\epsilon_n\}_{n=1}^{\infty}$ and $\{\delta_n\}_{n=1}^{\infty}$:

$$If \quad X_n = o_P\left(\epsilon_n\right) \quad and \quad Y_n = o_P\left(\epsilon_n\right), \quad then \quad X_n \pm Y_n = o_P\left(\epsilon_n\right). \tag{4.8}$$

$$If \quad X_n = o_P\left(\epsilon_n\right) \quad and \quad Y_n = O_P\left(\delta_n\right), \quad then \quad X_n Y_n = o_P\left(\delta_n \epsilon_n\right). \tag{4.9}$$

In particular, (4.9) implies (4.10).

$$If \quad X_n = o_P\left(\epsilon_n\right) \quad and \quad Y_n = o_P\left(\delta_n\right), \quad then \quad X_n Y_n = o_P\left(\delta_n \epsilon_n\right). \tag{4.10}$$

Moreover, for $a \in \mathbb{R}$ and a continuous function $g : \mathbb{R} \to \mathbb{R}$, we have

$$If \quad X_n \xrightarrow{P} a, \quad then \quad g(X_n) \xrightarrow{P} g(a). \tag{4.11}$$

Equations (4.8)–(4.10) are known as *Slutsky's theorem* and Eq. (4.11) is known as the *Continuous Mapping Theorem*. For proofs, see [16], for example. The notation O_P, o_P is not commonly used in general statistics, but it is frequently used in Watanabe's Bayesian theory, so it is necessary to understand it well.

Example 39 An independent sequence of random variables $X_1, X_2, \ldots$ such that $X_n \sim N(0, 1)$ is $O_P(1)$. Also, a sequence of random variables $X_1, X_2, \ldots$ such that $X_n \sim N(0, 1/n)$ stochastically converges to 0, hence $X_n = o_P(1)$. ∎

Proposition 6 (Weak Law of Large Numbers) *For a sequence of independent and identically distributed random variables $\{X_n\}$, the average $Z_n := \dfrac{X_1 + \cdots + X_n}{n}$ stochastically converges to its expected value*[5] μ.

Proof First, the mean and variance of Z_n are, respectively, $\mathbb{E}[Z_n] = \mathbb{E}[\dfrac{X_1 + \cdots + X_n}{n}] = \mathbb{E}[X_1] = \mu$ and[6] $\mathbb{V}[Z_n] = \mathbb{V}[\dfrac{X_1 + \cdots + X_n}{n}] = \mathbb{V}[X_1]/n = \sigma^2/n$. Applying these to Proposition 5, we obtain

---

[5] There is also a Strong Law of Large Numbers, which states that under the same conditions, there is almost sure convergence, not just convergence in probability, but this book does not deal with almost sure convergence.

[6] For $a, b \in \mathbb{R}$ and a random variable X, we have $\mathbb{E}[aX + b] = a\mathbb{E}[X] + b$, $\mathbb{V}[aX + b] = a^2\mathbb{V}[X]$.

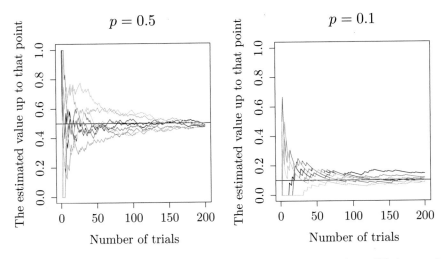

Fig. 4.2 A random number following a binomial distribution is generated $n = 200$ times, and the convergence of the sequence of random variables Z_n is illustrated. A sequence was generated eight times each for the probabilities of 1 occurring $p = 0.5$ and $p = 0.1$ (Example 40). Since the variances of X_n for $p = 0.5, 0.1$ are $p(1 - p) = 0.25, 0.09$ respectively, the variance at each $i = 1, \ldots, n$ of Z_i is $0.25/i, 0.09/i$. It can be seen that the estimated values up to that point are converging to $p = 0.5, 0.1$ respectively

$$P(|Z_n - \mu| \geq \epsilon) \leq \frac{\sigma^2}{n}/\epsilon^2.$$

Therefore, as $n \to \infty$, the probability of the event $(|Z_n - \mu| \geq \epsilon)$ approaches 0. ∎

Example 40 We generated random numbers following a binomial distribution 200 times ($n = 200$), calculated Z_i for each point up to $i = 1, \ldots, n$, and checked the degree of convergence (see Fig. 4.2). We generated Z_n 8 times each for $p = 0.5$ and $p = 0.1$.

```r
n <- 200
p <- 0.5
plot(0,0,xlim=c(1,n),ylim=c(0,1),xlab="Number oftrials",ylab="The
      estimated value up to that point",
  type="n",main="p=0.5")
abline(h=p)
for(j in 1:8){
  x <- rbinom(n,1,p)
  y <- NULL
  for(i in 1:n) y <- c(y,sum(x[1:i])/i)
  lines(1:n,y,col=j)
}
```

∎

4.3.4 Central Limit Theorem

In the following, we denote the mean and variance of the true distribution q as μ and σ^2, respectively. The *Central Limit Theorem* is, alongside the Law of Large Numbers, an important asymptotic property of a sequence of random variables $\{X_n\}$.

Proposition 7 (Central Limit Theorem) *For a sequence of independent random variables $\{X_n\}$ each following the same distribution with mean μ and variance σ^2,*

$$Y_n := \frac{X_1 + \cdots + X_n - n\mu}{\sigma\sqrt{n}} \tag{4.12}$$

follows the standard normal distribution as $n \to \infty$.

This book will not prove this theorem, but we will confirm its meaning by giving examples of this theorem and its extensions. First, it should be noted that each of the random variables in the sequence X_n does not necessarily need to follow a normal distribution.

Example 41 (*Application of the Central Limit Theorem*) Setting $n = 100$, for each distribution q below, we generated $m = 500$ random samples of (4.12) and plotted the distribution of Y_n (see Fig. 4.3).

1. Standard normal distribution
2. Exponential distribution with $\lambda = 1$
3. Binomial distribution with $p = 0.1$
4. Poisson distribution with $\lambda = 1$

Note that the exponential distribution is a distribution with a probability density function that is 0 for $x \le 0$ and

$$q(x) := \lambda e^{-\lambda x}$$

for $x \ge 0$. The Poisson distribution takes values $x = 0, 1, 2, \ldots$, with probabilities $q(x) = e^{-\lambda}\lambda^x / x!$. The experiment was run using the following code:

```
CLT <- function(dist){
  S <- NULL
  for(j in 1:500){
    if(dist==1) x <- rnorm(n,mu,sigma)
    if(dist==2) x <- rexp(n,lambda)
    if(dist==3) x <- rbinom(n,1,p)
    if(dist==4) x <- rpois(n,lambda)
    S <- c(S,(sum(x)-n*mu)/sqrt(n)/sigma)
  }
  titles <- c("Normal distribution","Exponential distribution (lambda=1)"
    ,"Binomial distribution (p=1)","Poisson distribution (lambda=1)")
  plot(density(S),xlab="Y_n",ylab="Probability density",main=titles[dist
    ])
}
```

```
15  ## Normal distribution
16  n=100
17  mu <- 0
18  sigma <- 1
19  CLT(1)
20  ## Exponential distribution
21  lambda <- 1
22  mu <- 1/lambda
23  sigma <- 1/lambda
24  CLT(2)
25  ## Binomial distribution
26  p <- 0.1
27  mu <- p
28  sigma <-sqrt(p*(1-p))
29  CLT(3)
30  ## Poisson distribution
31  lambda <- 1
32  mu <- lambda
33  sigma <- sqrt(lambda)
34  CLT(4)
35  ## Specify dist as 1 - 4, and set the parameters for each distribution as
36       above
```

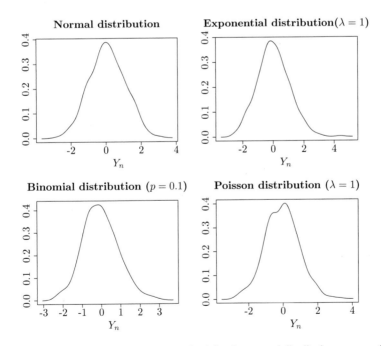

Fig. 4.3 We generated $n = 100$ random samples following normal distribution, exponential distribution, binomial distribution, and Poisson distribution, and calculated the value of Y_n once. This process was repeated $m = 500$ times to examine its distribution. Even with $n = 100$, the distribution is approaching a normal distribution

It can be seen that regardless of the shape of distribution q, even with $n = 100$, the shape is close to the standard normal distribution. ∎

The above Central Limit Theorem assumed that $X_1, \ldots, X_n$ were each real numbers (one-dimensional), and assumed $\mu \in \mathbb{R}$, $\sigma^2 > 0$. Similar assertions hold even for two-dimensional and d-dimensional ($d \geq 1$) cases. Hereinafter, $N(\mu, \Sigma)$ denotes a d-dimensional normal distribution with mean $\mu \in \mathbb{R}^d$ and covariance matrix $\Sigma \in \mathbb{R}^{d \times d}$. The probability density function of $X \sim N(\mu, \Sigma)$ is as follows.

$$f(x) = \frac{1}{(2\pi)^{d/2}(\det \Sigma)^{1/2}} \exp\{-\frac{1}{2}(x - \mu)^\top \Sigma^{-1}(x - \mu)\}.$$

In general, when the distribution function $\{F_n(x)\}$ of a sequence of real-valued random variables $\{X_n\}$ converges to the distribution function $F_X(x) := \int_{-\infty}^x q(t)dt$ of a random variable X at each continuous point x as $n \to \infty$,

$$\lim_{n \to \infty} F_n(x) = F_X(x) \tag{4.13}$$

we say that $\{X_n\}$ converges in distribution to X, and write this as $X_n \xrightarrow{d} X$. If the probability density function followed by X is q, we sometimes write this as $X_n \xrightarrow{d} q$. For example, the Central Limit Theorem can be written as $X_n \xrightarrow{d} N(0, 1)$. And, it is known that (4.13) is equivalent to

$$\lim_{n \to \infty} \mathbb{E}_n[g(X_n)] = \mathbb{E}_X[g(X)] \tag{4.14}$$

for any bounded and continuous function $g : \mathbb{R} \to \mathbb{R}$ (Exercise 38), where $\mathbb{E}_n[\cdot]$, $\mathbb{E}_X[\cdot]$ are the operations of the mean with respect to the distribution functions F_n, F_X respectively.

Proposition 8 *Consider independent random variables $X_1, \ldots, X_n$ with mean $\mu \in \mathbb{R}^d$ and covariance matrix $\Sigma \in \mathbb{R}^{d \times d}$ (they do not necessarily follow a normal distribution). Then, we have*

$$\frac{X_1 + \cdots + X_n - n\mu}{\sqrt{n}} \xrightarrow{d} N(0, \Sigma).$$

On the other hand, for a probability variable $\eta_n : C(K) \to \mathbb{R}$ that takes values in $C(K)$, the concept of distribution function does not exist because $C(K)$ is not in a Euclidean space.[7] Therefore, for any bounded and continuous function $g : C(K) \to \mathbb{R}$, we define the convergence in distribution of the sequence $\eta_1, \eta_2, \ldots$ to a random variable η taking values in some $C(K)$ ($\eta_n \xrightarrow{d} \eta$) as

---

[7] The concept of distribution function applies not only to the one-dimensional case $F : \mathbb{R} \ni x \mapsto \int_{-\infty}^x q(t)dt$, but also to the two-dimensional case $F : \mathbb{R}^2 \ni (x, y) \mapsto \int_{-\infty}^x \int_{-\infty}^y q(s, t)ds dt$.

$$\lim_{n \to \infty} \mathbb{E}_n[g(\eta_n)] = \mathbb{E}_\eta[g(\eta)]. \tag{4.15}$$

4.4 Fisher Information Matrix

The *Fisher information matrix* represents the smoothness of the log-likelihood $\log p(X|\theta)$ at each $\theta \in \Theta$, and is an important measure for analyzing the relationship between the true distribution and the statistical model.

In this book, we assume the following conditions.

Assumption 2 1. The order of integration in $\mathcal{X}$ and differentiation with respect to $\theta \in \Theta$ in $p(\cdot|\theta)$ can be exchanged.

2. For each $(x, \theta) \in \mathcal{X} \times \Theta$, the partial derivatives $\dfrac{\partial^2 \log p(x|\theta)}{\partial \theta_i \partial \theta_j}$ exist, for $i, j = 1, \ldots, d$.

The Fisher information matrix $I(\theta)$ is defined as the covariance matrix of

$$\nabla \log p(X|\theta) = \left[\frac{\partial \log p(X|\theta)}{\partial \theta_1}, \ldots, \frac{\partial \log p(X|\theta)}{\partial \theta_d} \right]$$

$$
\begin{aligned}
I(\theta) &:= \mathbb{V}[\nabla \log p(X|\theta)] \\
&= \mathbb{E}_X[\{\nabla \log p(X|\theta) - \mathbb{E}_{X'}[\nabla \log p(X'|\theta)]\} \\
&\quad \cdot \{\nabla \log p(X|\theta) - \mathbb{E}_{X''}[\nabla \log p(X''|\theta)]\}^\top] \\
&= \mathbb{E}_X[\nabla \log p(X|\theta)(\nabla \log p(X|\theta))^\top] - \mathbb{V}\mathbb{E}_X[\log p(X|\theta)]\nabla \mathbb{E}_X[\log p(X|\theta)]^\top
\end{aligned}
\tag{4.16}
$$

and we denote $I := I(\theta_*) \in \mathbb{R}^{d \times d}$ for $\theta_* \in \Theta_*$. Also, we define the matrix $J := J(\theta_*) \in \mathbb{R}^{d \times d}$ using

$$J(\theta) := \mathbb{E}_X[-\nabla^2 \log p(X|\theta)]. \tag{4.17}$$

Assuming regularity, there exists a unique $\theta = \theta_*$ that minimizes $D(q\|p(\cdot|\theta))$, that is, minimizes $\mathbb{E}_X[-\log p(X|\theta)]$, and there exists an open set containing θ_* that is included in Θ, and since $\nabla^2 \mathbb{E}_X[\log p(X|\theta)]$ is positive definite, $\nabla \mathbb{E}_X[\log p(X|\theta)]$ is 0 at $\theta = \theta_*$. We write this as

$$\nabla \mathbb{E}_X[\log p(X|\theta_*)] = 0. \tag{4.18}$$

Therefore, if it is regular, the following holds from (4.16).

$$I(\theta_*) = \mathbb{E}_X[\nabla \log p(X|\theta_*)(\nabla \log p(X|\theta_*))^\top]. \tag{4.19}$$

Example 42 Assume that the mean and variance of the true distribution q are μ and σ^2, respectively (not necessarily normally distributed). For the probability density function (normal distribution) with parameter $\theta = (\mu, \sigma^2)$

$$p(x|\theta) = \frac{1}{\sqrt{2\pi\sigma^2}} \exp -\frac{(x-\mu)^2}{2\sigma^2},$$

we shall calculate the matrices I, J. From

$$\log p(x|\theta) = -\frac{1}{2} \log 2\pi\sigma^2 - \frac{(x-\mu)^2}{2\sigma^2}$$

$$\nabla[\log p(x|\theta)] = \begin{bmatrix} \dfrac{x-\mu}{\sigma^2} \\ -\dfrac{1}{2\sigma^2} + \dfrac{(x-\mu)^2}{2(\sigma^2)^2} \end{bmatrix} = \begin{bmatrix} \dfrac{x-\mu}{\sigma^2} \\ \dfrac{(x-\mu)^2 - \sigma^2}{2(\sigma^2)^2} \end{bmatrix} \qquad (4.20)$$

$$\nabla^2[\log p(x|\theta)] = \begin{bmatrix} -\dfrac{1}{\sigma^2} & -\dfrac{x-\mu}{(\sigma^2)^2} \\ -\dfrac{x-\mu}{(\sigma^2)^2} & \dfrac{1}{2(\sigma^2)^2} - \dfrac{(x-\mu)^2}{(\sigma^2)^3} \end{bmatrix} \qquad (4.21)$$

$$\mathbb{E}_X[(X-\mu)^2] = \mathbb{E}_X[(X - \mu_{**} + \mu_{**} - \mu)^2] = \sigma_{**}^2 + (\mu_{**} - \mu)^2, \qquad (4.22)$$

we obtain

$$\mathbb{E}_X[\nabla \log p(X|\theta)] = \begin{bmatrix} \dfrac{\mu_{**} - \mu}{\sigma^2} \\ -\dfrac{1}{2\sigma^2} + \dfrac{\sigma_{**}^2 + (\mu_{**} - \mu)^2}{2(\sigma^2)^2} \end{bmatrix} \qquad (4.23)$$

$$\mathbb{V}[\nabla \log p(X|\theta)]$$
$$= \mathbb{E}_X[\{\nabla \log p(X|\theta) - \mathbb{E}_X[\nabla \log p(X|\theta)]\} \{\nabla \log p(X|\theta) - \mathbb{E}_X[\nabla \log p(X|\theta)]\}^\top]$$
$$= \mathbb{E}_X\{ \begin{bmatrix} \dfrac{X - \mu_{**}}{\sigma^2} \\ \dfrac{(X - \mu_{**})^2 + 2(\mu_{**} - \mu)(X - \mu_{**}) - \sigma_{**}^2}{2(\sigma^2)^2} \end{bmatrix}$$
$$\cdot \begin{bmatrix} \dfrac{X - \mu_{**}}{\sigma^2} & \dfrac{(X - \mu_{**})^2 + 2(\mu_{**} - \mu)(X - \mu_{**}) - \sigma_{**}^2}{2(\sigma^2)^2} \end{bmatrix} \}. \qquad (4.24)$$

Let $A := \mathbb{E}_X[(X - \mu_{**})^3]$ and $B := \mathbb{E}_X[(X - \mu_{**})^4]$, then the (1,1), (1,2), and (2,2) elements of (4.24) are respectively

$$\mathbb{E}_X[\left(\frac{X - \mu_{**}}{\sigma^2} \right)^2] = \frac{\sigma_{**}^2}{(\sigma^2)^2},$$

$$\mathbb{E}_X\left[\frac{X - \mu_{**}}{\sigma^2} \cdot \frac{(X - \mu_{**})^2 + 2(\mu_{**} - \mu)(X - \mu_{**}) - \sigma_{**}^2}{2(\sigma^2)^2}\right] = \frac{A + 2(\mu_{**} - \mu)\sigma_{**}^2}{2(\sigma^2)^3},$$

and

$$\mathbb{E}_X\left[\left\{\frac{(X - \mu_{**})^2 + 2(\mu_{**} - \mu)(X - \mu_{**}) - \sigma_{**}^2}{2(\sigma^2)^2}\right\}^2\right]$$

$$= \frac{1}{4(\sigma^2)^4}\left\{\mathbb{E}_X[(X - \mu_{**})^4] + 4(\mu_{**} - \mu)\mathbb{E}_X[(X - \mu_{**})^3]\right.$$

$$\left. +4(\mu_{**} - \mu)^2\mathbb{E}_X[(X - \mu_{**})^2] + (\sigma_{**}^2)^2 - 2\sigma_{**}^2\mathbb{E}_X[(X - \mu_{**})^2]\right\}$$

$$= \frac{B - (\sigma_{**}^2)^2 + 4(\mu_{**} - \mu)A + 4(\mu_{**} - \mu)^2\sigma_{**}^2}{4(\sigma^2)^4}.$$

Furthermore, by substituting $\theta = \theta_* = (\mu_*, \sigma_*^2)$, (4.24) becomes as follows.

$$I = \begin{bmatrix} \dfrac{\sigma_{**}^2}{(\sigma_*^2)^2} & \dfrac{A + 2(\mu_{**} - \mu_*)\sigma_{**}^2}{2(\sigma_*^2)^3} \\ \dfrac{A + 2(\mu_{**} - \mu_*)\sigma_{**}^2}{2(\sigma_*^2)^3} & \dfrac{B - (\sigma_{**}^2)^2 + 4(\mu_{**} - \mu_*)A + 4(\mu_{**} - \mu_*)^2\sigma_{**}^2}{4(\sigma_*^2)^4} \end{bmatrix}. \tag{4.25}$$

On the other hand, from (4.21), we obtain

$$J(\theta) = \begin{bmatrix} \dfrac{1}{\sigma^2} & \dfrac{\mu_{**} - \mu}{(\sigma^2)^2} \\ \dfrac{\mu_{**} - \mu}{(\sigma^2)^2} & -\dfrac{1}{2(\sigma^2)^2} + \dfrac{\sigma_{**}^2 + (\mu_{**} - \mu)^2}{(\sigma^2)^3} \end{bmatrix}$$

and

$$J = \begin{bmatrix} \dfrac{1}{\sigma_*^2} & \dfrac{\mu_{**} - \mu_*}{(\sigma_*^2)^2} \\ \dfrac{\mu_{**} - \mu_*}{(\sigma_*^2)^2} & -\dfrac{1}{2(\sigma_*^2)^2} + \dfrac{\sigma_{**}^2 + (\mu_{**} - \mu_*)^2}{(\sigma_*^2)^3} \end{bmatrix}. \tag{4.26}$$

Moreover, if it is regular, from (4.19), (4.23) becomes 0, so $(\mu, \sigma_*^2) = (\mu, \sigma^2)$. Therefore, we obtain

$$I = \begin{bmatrix} (\sigma_*^2)^{-1} & \dfrac{A}{2(\sigma_*^2)^3} \\ \dfrac{A}{2(\sigma_*^2)^3} & \dfrac{B - (\sigma_*^2)^2}{4(\sigma_*^2)^4} \end{bmatrix} \tag{4.27}$$

and

$$J = \begin{bmatrix} (\sigma_*^2)^{-1} & 0 \\ 0 & (\sigma_*^2)^{-2}/2 \end{bmatrix}. \tag{4.28}$$

Furthermore, if it is realizable, the true distribution q is also normal, and since $A = 0$ and $B = 3(\sigma_{**}^2)^2$ (as per Example 2), (4.27) coincides with (4.28). ∎

Proposition 9 *When the true distribution q is realizable for the statistical model $p(\cdot|\theta)_{\theta \in \Theta}$ and is regular, $I = J$ holds.*

Proof Since it is realizable, $q = p(\cdot|\theta_*)$, and we can write

$$
\begin{aligned}
J &= \mathbb{E}_X[-\nabla^2 \log p(X|\theta_*)] = \mathbb{E}_X\left[-\nabla\left(\frac{\nabla p(X|\theta_*)}{p(X|\theta_*)}\right)\right] \\
&= \mathbb{E}_X\left[-\frac{\nabla^2 p(X|\theta_*)}{p(X|\theta_*)}\right] + \mathbb{E}_X\left[\frac{\nabla p(X|\theta_*)(\nabla p(X|\theta_*))^\top}{p(X|\theta_*)^2}\right] \\
&= -\mathbb{E}_X\left[\frac{\nabla^2 p(X|\theta_*)}{q(X)}\right] + \mathbb{E}_X\left[\nabla \log p(X|\theta_*)(\nabla \log p(X|\theta_*))^\top\right].
\end{aligned}
$$

Furthermore, from the first condition of Assumption 2, we have

$$
\mathbb{E}_X[\frac{\nabla^2 p(X|\theta_*)}{q(X)}] = \int_{\mathcal{X}} \nabla^2 p(x|\theta_*)dx = \nabla^2 \int_{\mathcal{X}} p(x|\theta_*)dx = \nabla^2 1 = 0,
$$

and from the equation where we substitute $\theta = \theta_*$ into (4.16), we can write

$$
J = 0 + I(\theta_*) + (\nabla\mathbb{E}_X[-\log p(X|\theta_*)]) (\nabla\mathbb{E}_X[-\log p(X|\theta_*)])^\top.
$$

Furthermore, since it is regular, we can apply (4.18), and the proposition follows. ∎

Example 43 In Example 42, if we change the parameter $\sigma^2 > 0$ to $\sigma \neq 0$, the distribution becomes the same (homogeneous) at $\theta = (\mu, \sigma)$ and $\theta = (\mu, -\sigma)$, but

$$
\nabla[\log p(x|\theta)] = [\frac{x - \mu}{\sigma^2}, -\frac{1}{\sigma} + \frac{(x - \mu)^2}{\sigma^3}]^\top,
$$

$$
\nabla^2[\log p(x|\theta)] = \begin{bmatrix} -\dfrac{1}{\sigma^2} & -\dfrac{2(x - \mu)}{\sigma^3} \\ -\dfrac{2(x - \mu)}{\sigma^3} & \dfrac{1}{\sigma^2} - \dfrac{3(x - \mu)^2}{\sigma^4} \end{bmatrix},
$$

and

$$
J(\theta) = \begin{bmatrix} \dfrac{1}{\sigma^2} & \dfrac{2(\mu_{**} - \mu)}{\sigma^3} \\ \dfrac{2(\mu_{**} - \mu)}{\sigma^3} & -\dfrac{1}{\sigma^2} + 3\dfrac{(\mu_{**} - \mu)^2 + \sigma_{**}^2}{\sigma^4} \end{bmatrix},
$$

follow, and the values of $J(\theta)$ do not coincide between the two. When $\mu = \mu_{**}$, they coincide at $\pm\sigma$. ∎

When limited to the exponential family, using the notation of Sect. 2.4, given that $p(x|\theta) = u(x)\exp\{v(\theta)^\top w(x)\}$ and $\nabla \log p(x|\theta) = \nabla v(\theta)^\top w(x)$, the Fisher information matrix can be written as

$$I(\theta) = \mathbb{V}[\nabla\{v(\theta)^\top w(X)\}]$$

and

$$J(\theta) = -\mathbb{E}_X[\nabla^2\{v(\theta)^\top w(X)\}].$$

Example 44 In Example 42, with $J = 4$, we can write $u(x) = \dfrac{1}{\sqrt{2\pi}}$, $v(\theta) = [\dfrac{1}{\sigma^2}, \dfrac{\mu}{\sigma^2}, \dfrac{\mu^2}{\sigma^2}, \log\sigma^2]^\top$, and $w(x) = [-\dfrac{x^2}{2}, x, -\dfrac{1}{2}, -\dfrac{1}{2}]^\top$ (see Example 16). Hence,

$$\nabla[v(\theta)^\top w(x)] = \nabla[-\frac{1}{2}\log\sigma^2 - \frac{(x-\mu)^2}{2\sigma^2}] = [\frac{x-\mu}{\sigma^2}, -\frac{1}{2\sigma^2} + \frac{(x-\mu)^2}{2(\sigma^2)^2}]^\top$$

$$\nabla^2[v(\theta)^\top w(x)] = \begin{bmatrix} -\dfrac{1}{\sigma^2} & -\dfrac{x-\mu}{(\sigma^2)^2} \\[2mm] -\dfrac{x-\mu}{(\sigma^2)^2} & \dfrac{1}{2(\sigma^2)^2} - \dfrac{(x-\mu)^2}{(\sigma^2)^3} \end{bmatrix}$$

can be obtained. We can calculate $\mathbb{E}_X[\cdot]$ and $\mathbb{V}_X[\cdot]$ using these and the true model.
∎

Appendix: Proof of Proposition

Proof of Proposition 4

When matrix A is symmetric, the eigenvectors corresponding to different eigenvalues are orthogonal (their inner product is 0). Indeed, if $Au = \lambda u$, $Au' = \lambda'u'$, and $\lambda \neq \lambda'$, we have

$$0 = \langle Au, u'\rangle - \langle Au, u'\rangle = \langle Au, u'\rangle - \langle u, Au'\rangle = \langle \lambda u, u'\rangle - \langle u, \lambda'u'\rangle = (\lambda - \lambda')\langle u, u'\rangle$$

which results in $\langle u, u'\rangle = 0$. If the eigenvalues are repeated, there exist as many linearly independent eigenvectors as there are repetitions, and we choose them to be orthogonal. Moreover, we normalize all eigenvectors to have a magnitude of 1. Assume that we have obtained eigenvalues $\lambda_1, \ldots, \lambda_n$ and eigenvectors $u_1, \ldots, u_n$ in this way. In this case, $Au_i = \lambda_i u_i$ holds. Also, let $U \in \mathbb{R}^{n\times n}$ be the matrix with columns $u_1, \ldots, u_n$, and let $D \in \mathbb{R}^{n \times n}$ be the diagonal matrix with diagonal entries $\lambda_1, \ldots, \lambda_n$. We then have $AU = UD$. Since U is an orthogonal matrix due to its construction ($U^\top U = U^\top U = I_n$), we can multiply $U^\top$ from the right to get $A = UDU^\top$.

First, if $z^\top A z \geq 0$ for any $z \in \mathbb{R}^n$, then $\lambda_i u_i^\top u_i \geq 0$ and $\lambda_i \geq 0$. Conversely, if $\lambda_i \geq 0$, by denoting the matrix obtained by replacing each component of D with its square root as $\sqrt{D}$, we have $A = U\sqrt{D}\sqrt{D}U^\top = (\sqrt{D}U^\top)^\top \sqrt{D}U^\top$, and $z^\top A z = (\sqrt{D}U^\top z)^\top \sqrt{D}U^\top z \geq 0$ holds for any $z \in \mathbb{R}^n$.

Furthermore, if $z^\top A z > 0$ for any $z \neq 0$, then $\lambda_i u_i^\top u_i > 0$ and, since $u_i^\top u_i \neq 0$, we have $\lambda_i > 0$. Conversely, if $\lambda_i > 0$, then $\sqrt{D}U^\top$ is regular, and for any $z \neq 0$, $\sqrt{D}U^\top z \neq 0$, and $z^\top A z = (\sqrt{D}U^\top z)^\top \sqrt{D}U^\top z \neq 0$ holds. ∎

Proof of Proposition 2

In the following, let $f(x, \theta_*, \theta) := \log \frac{p(x|\theta_*)}{p(x|\theta)}$.

1. Given any $\theta_1, \theta_2 \in \Theta_*$, we have relative finite variance, so

$$0 = D(q \| p(\cdot|\theta_2)) - D(q \| p(\cdot|\theta_1)) = \int_{\mathcal{X}} q(x) f(x, \theta_1, \theta_2) dx \geq \gamma \int_{\mathcal{X}} q(x) f(x, \theta_1, \theta_2)^2 dx \geq 0.$$

There exists a constant $\gamma > 0$, so $f(\cdot, \theta_1, \theta_2)$ is zero as a function, and θ_1, θ_2 become the same distribution.

2. As q is realizable, we can set $f(x, \theta_*, \theta) = \log \frac{q(x)}{p(x|\theta)}$. Arbitrarily choose $\theta_* \in \Theta_*$ from the homogeneous Θ_* and consider the limit of $\mathbb{E}_X[f(X, \theta_*, \theta)] = D(q|p(\cdot|\theta))$ as $\theta \to \theta_*$ in (2.22). For $F(t) := t + e^{-t} - 1$, $t \in \mathbb{R}$, there exists $|t^*| \leq |t|$ such that $F(t) = \frac{t^2}{2} e^{-t^*}$ from the Taylor expansion of F at $t = 0$. Here, note that

$$F(\log \frac{q(x)}{p(x|\theta)}) = \log \frac{q(x)}{p(x|\theta)} + \frac{p(x|\theta)}{q(x)} - 1.$$

Then, we can see that

$$\mathbb{E}_X[f(X, \theta_*, \theta)] = \int_{\mathcal{X}} q(x) F(\log \frac{q(x)}{p(x|\theta)}) dx = \frac{1}{2} \int_{\mathcal{X}} f(x, \theta_*, \theta)^2 e^{-t_\theta(x)} q(x) dx$$

$$\geq \frac{1}{2} \int_{\mathcal{X}} f(x, \theta_*, \theta)^2 \min\{\frac{p(x|\theta)}{q(x)}, \frac{q(x)}{p(x|\theta)}\} q(x) dx$$

holds, where we assume $|t_\theta(x)| \leq |\log \frac{q(x)}{p(x|\theta)}|$. Also, from the continuity (Assumption 1), if $\theta \to \theta_*$, we can make $p(x|\theta) \to q(x) = p(x|\theta_*)$. Therefore, for any $\epsilon > 0$, there exists a θ that satisfies

$$\min\{\frac{p(x|\theta)}{q(x)}, \frac{q(x)}{p(x|\theta)}\} > \frac{1}{1+\epsilon}.$$

Hence, if $|\theta - \theta_*| < \delta$, we can have

$$\mathbb{E}_X[f(X, \theta_*, \theta)] \geq \frac{1}{2(1 + \epsilon)} \mathbb{E}_X[f(X, \theta_*, \theta)^2]$$

and the constant c in (2.22) can be bounded by 2.

3. Let $g(\theta) := \mathbb{E}_X[f(X, \theta_*, \theta)]$ and $h(\theta) := \mathbb{E}_X[f^2(X, \theta_*, \theta)]$. Both $g(\theta)$ and $h(\theta)$ take their minimum values $g(\theta_*) = h(\theta_*) = 0$ at θ_*, an element of Θ_*. Hence, $\nabla g(\theta_*) = \nabla h(\theta_*) = 0$. Therefore, by Taylor's expansion, and defining $\nabla g(\theta) = \left(\frac{\partial g(\theta)}{\partial \theta_1} |_{\theta=\theta_*}, \ldots, \frac{\partial g(\theta)}{\partial \theta_d} |_{\theta=\theta_*} \right)$, we have

$$g(\theta) = g(\theta_*) + \nabla g(\theta_*)^\top (\theta - \theta_*) + \frac{1}{2}(\theta - \theta_*)^\top \nabla^2 g(\theta_1)(\theta - \theta_*) = \frac{1}{2}(\theta - \theta_*)^\top \nabla^2 g(\theta_1)(\theta - \theta_*)$$

and

$$h(\theta) = h(\theta_*) + \nabla h(\theta_*)^\top (\theta - \theta_*) + \frac{1}{2}(\theta - \theta_*)^\top \nabla^2 h(\theta_2)(\theta - \theta_*) = \frac{1}{2}(\theta - \theta_*)^\top \nabla^2 h(\theta_2)(\theta - \theta_*)$$

for some θ_1, θ_2 that exist between θ and θ_*. Moreover, as $\theta \to \theta_*$, we have $\theta_1, \theta_2 \to \theta_*$, and both can be approximated in the neighborhood of θ_* by

$$\frac{1}{2}(\theta - \theta_*)^\top \nabla^2 g(\theta_*)(\theta - \theta_*) , \quad \frac{1}{2}(\theta - \theta_*)^\top \nabla^2 h(\theta_*)(\theta - \theta_*).$$

On the other hand, since g is regular, the θ_* that minimizes g is unique, and all eigenvalues of $\nabla^2 g(\theta_*) = \nabla^2 D(q||p(\cdot|\theta))|_{\theta=\theta_*}$ are positive. Also, according to Proposition 4, $\nabla^2 h$ is non-negative in the neighborhood of $\theta = \theta_*$. If we denote the smallest eigenvalue of the former as $\lambda_{min} > 0$ and the largest value of the latter as $\lambda_{max} \geq 0$, then we can write

$$\frac{\mathbb{E}_X[f(X, \theta_*, \theta)^2]}{\mathbb{E}_X[f(X, \theta_*, \theta)]} = \frac{h(\theta)}{g(\theta)} \leq \frac{\lambda_{max}}{\lambda_{min}}.$$

Exercises 27–41

27. For $a, b, c, d \in \mathbb{R}$, prove that $\overline{(a + bi)(c + di)} = \overline{a + bi} \cdot \overline{c + di}$. Also, for the eigenvalue $\lambda \in \mathbb{C}$ and eigenvector $u \in \mathbb{C}^n$ of matrix $A \in \mathbb{R}^{n \times n}$, prove that $A\overline{u} = \overline{Au} = \overline{\lambda u} = \overline{\lambda}\overline{u}$.

28. For a matrix $U = [u_1, \ldots, u_n] \in \mathbb{R}^{n \times n}$, where the inner product $\langle u_i, u_j \rangle$ of each column is 1 when $i = j$ and 0 otherwise, we call U an orthogonal matrix. Show that $U^\top U = UU^\top = I_n$.

29. Prove the following.

(a) An open interval (a, b) is an open set, and a closed interval $[a, b]$ is a closed set.

(b) The set of all real numbers $\mathbb{R}$ and the set of all integers $\mathbb{Z}$ are closed sets.

(c) The set $\mathbb{R} \cap \mathbb{Z}^C$, which is the set of all real numbers $\mathbb{R}$ excluding the set of all integers $\mathbb{Z}$, is an open set.

(d) The set of all rational numbers $\mathbb{Q}$ is neither an open set nor a closed set.

(e) The region $\{(x, y, z) \in \mathbb{R}^3 \mid x^2 + y^2 + z^2 < 1, z \geq 0\}$ is neither an open set nor a closed set.

30. Based on the definition of Maclaurin expansion, prove (4.4) (4.5).

31. Show that the function f in Example 37 is C^∞.

32. Show that the absolute value $|\cdot|$ is a norm in $\mathbb{R}$. Also, show that $L^2(q)$ is a linear space, and that $\|\cdot\|_2$ is a norm. [Hint] Show in relation to the relationship $\sim$ such as $f \sim g \iff \int_\mathcal{X} f(x)q(x)dx = \int_\mathcal{X} g(x)q(x)dx$.

33. When tossing a coin with equal probability of heads or tails, what kind of event set should be prepared for the variable X to become a random variable, with $X = 1$ if heads appear and $X = 0$ if tails appear?

34. Prove the two inequalities

$$\mathbb{E}[(X - \mu)^2] \geq \mathbb{E}[(X - \mu)^2 I(|X - \mu| \geq k)] \geq k^2 \cdot P(|X - \mu| \geq k).$$

35. Toss a coin with equal probability of heads or tails n times, and let a_i be the relative frequency of heads occurring up to the ith time, for $1 \leq i \leq n$. Write an R program that takes n as input, generates n random numbers following a binomial distribution, and outputs the sequence $a_1, \ldots, a_n$.

36. When executing the following program with different values of m and n such as $m = 10, 100$ and $n = 10, 100$, different graphs are obtained. What kind of graphs can be generally obtained?

```
1  m <- 100
2  n <- 100
3  x.seq <- NULL
4  for (i in 1:n) {
5    x <- (sum(rbinom(m, 1, 0.5)) - 0.5*m) / (sqrt (m/4))
6    x.seq <- c(x.seq, x)
7  }
8  curve(dnorm(x), -5, 5, col=2)
9  lines(density(x.seq), col=3)
```

37. Following the application example of Example 41, generate $n = 500$ random numbers approximately following the standard normal distribution from $m = 100$ sets of random numbers following the χ^2 distribution with 2 degrees of freedom, and draw a graph similar to Fig. 4.3.

38. Prove the following two propositions.

(a) If $X_n \xrightarrow{d} X$ and $g : \mathbb{R} \to \mathbb{R}$ is bounded ($|g(x)| < M$, there exists $M > 0$ such that $x \in \mathbb{R}$) and continuous, then $\mathbb{E}[g(X_n)] \to \mathbb{E}[g(X)]$, where the following fact can be used without proof. When fixing $\epsilon > 0$ arbitrarily, we can choose continuous points $a_0 < a_1 < \cdots < a_k$ of the distribution function of X that satisfy the following conditions:

$$P\ (X \le a_0) < \epsilon\ , \quad P\ (X > a_k) < \epsilon$$

and

$$|g(x) - g(a_i)| < \epsilon\ , \quad x \in \left[a_{i-1}, a_i\right]\ , \quad i = 1, \ldots, k.$$

[Hint] Apply the function $h : \mathbb{R} \to \mathbb{R}$, which takes the value 0 outside $(a_0, a_k]$ and a constant value within each (a_{i-1}, a_i), to the following inequality.

$$|\mathbb{E}[g(X_n)] - \mathbb{E}[g(X)]|$$
$$\le |\mathbb{E}[g(X_n)] - \mathbb{E}[h(X_n)]| + |\mathbb{E}[h(X_n)] - \mathbb{E}[h(X)]| + |\mathbb{E}[h(X)] - \mathbb{E}[g(X)]|.$$

(b) For any bounded and continuous $g : \mathbb{R} \to \mathbb{R}$, if $\mathbb{E}[g(X_n)] \to \mathbb{E}[g(X)]$, then $X_n \overset{d}{\to} X$. [Hint] Since $g : \mathbb{R} \to \mathbb{R}$ is a bounded and continuous function, for example, let $a \in \mathbb{R}$ be a continuous point of the distribution function of X, and let $m \ge 1$, then

$$g_{a,m}(x) = \begin{cases} 1, & x \le a \\ -m(x - a) + 1, & a < x < a + 1/m \\ 0, & x \ge a + 1/m. \end{cases}$$

For the function $g_{a,m} : \mathbb{R} \to \mathbb{R}$, $\mathbb{E}[g_{a,m}(X_n)] \to \mathbb{E}[g_{a,m}(X)]$ $(n \to \infty)$ is established, and $F_X(a) \le \mathbb{E}[g_{a,m}(X)] \le F_X(a + \frac{1}{m})$ holds. Finally, use the fact that $a \in \mathbb{R}$ is a continuous point of the distribution function.

39. When the true distribution is regular with respect to the statistical model, show that for $\theta_* \in \Theta_*$,

$$I(\theta_*) = \mathbb{E}_X[\nabla \log p(X|\theta_*)(\nabla \log p(X|\theta_*))^\top]$$

holds.

40. Under regularity, (4.23) becomes 0. Assuming $\theta = (\mu, \sigma^2) = (\mu_*, \sigma_*^2) = \theta_*$, show that (4.25) can be written by (4.27). Also, why does (4.27) become (4.28) when realizable?

41. Perform the same derivation as in Example 44 for Example 17.

Chapter 5
Regular Statistical Models

In this chapter, we discuss the situation where the true distribution has a regular relationship with the statistical model. We will explain the traditional approach before the emergence of Watanabe's Bayesian theory. Being regular, Θ_* contains a single element θ_*. In Watanabe's Bayesian theory, this is divided into Θ within a Euclidean distance of $\epsilon_n = n^{-1/4}$ (where n is the sample size) from θ_* and everything else. For the latter, we apply the discussion without assuming regularity. In other words, the generalization proposed by Watanabe's Bayesian theory applies only to the former. In this chapter, we will demonstrate existing analytical methods for the former, as well as explicitly show results that can also be applied in Chap. 8. First, we obtain a result regarding the posterior distribution (asymptotic normality). Then, we define important statistical quantities for defining WAIC, namely the generalization loss and empirical loss. Finally, we define specific values assuming regularity.

Relationship between Propositions of Chapter 5 and Regularity

| P.10 | ← | Regular | → | P.14 | → | P.15 | → | P.16 | → | P.19 |
| P.11 | → | P.12 | → | P.13 | | P.17 | → | P.18 | | |

(with ↑ above P.14, and ↗ from P.18 to P.19)

(Red is applicable only in Chaps. 5 and 6, blue can also be applied in Chaps. 8 and 9)

In this chapter, we assume that Θ is a compact set.

Assumption 3 Θ is compact

Note that, from this chapter onwards, there will be situations where we analyze the stochastic fluctuations of the training data, but unless we are taking the average, we will describe them in lowercase $(x_1, \ldots, x_n \in \mathcal{X})$ rather than uppercase $(X_1, \ldots, X_n \in \mathcal{X})$. Also, we write the marginal likelihood as Z_n rather than $Z(x_1, \ldots, x_n)$.

© The Author(s), under exclusive license to Springer Nature Singapore Pte Ltd. 2023
J. Suzuki, *WAIC and WBIC with R Stan*,
https://doi.org/10.1007/978-981-99-3838-4_5

5.1 Empirical Process

In this section, we will examine the properties of the following quantities. Let $J(\theta) \in \mathbb{R}^{d \times d}$ be the matrix defined in Sect. 4.4. For $x_1, \ldots, x_n \in \mathcal{X}, \theta \in \Theta$ and $\theta_* \in \Theta_*$, we define

$$\eta_n(\theta) := \frac{1}{\sqrt{n}} \sum_{i=1}^{n} \left\{ \mathbb{E}_X \left[\log \frac{p(X|\theta_*)}{p(X|\theta)} \right] - \log \frac{p(x_i|\theta_*)}{p(x_i|\theta)} \right\} \tag{5.1}$$

and

$$\Delta_n := J^{-1}(\theta_*) \nabla \eta_n(\theta_*) / \sqrt{n}. \tag{5.2}$$

where we assume the existence of the inverse matrix of $J(\theta_*)$, but η_n can be defined even if it is not regular. Applying the Central Limit Theorem, the following proposition holds.

Proposition 10 When $\theta \in \Theta$ is fixed, $\nabla \eta_n(\theta)$ and $\sqrt{n}\Delta_n$ converge in distribution to the normal distributions $N(0, I(\theta))$ and $N(0, J^{-1}IJ^{-1})$, respectively, and $\nabla \eta_n(\theta)/\sqrt{n}$ and Δ_n both converge in probability to 0, where $I = I(\theta_*)$ and $J = J(\theta_*)$.

Proof $\nabla \eta_n(\theta)$ is a value obtained by adding all independent $-\nabla \log p(x_i|\theta) + \mathbb{E}_X[\nabla \log p(X|\theta)]$ and dividing by $\sqrt{n}$. Also, its mean is 0, and from the Central Limit Theorem, it converges in distribution to a normal distribution with mean 0 and covariance matrix (4.16). Therefore, $\sqrt{n}\Delta_n = J^{-1}\nabla \eta_n$ converges in distribution to the normal distribution $N(0, J^{-1}IJ^{-1})$. Furthermore, that $\nabla \eta_n(\theta)/\sqrt{n}$ and Δ_n converge in probability to 0 follows from the Weak Law of Large Numbers. ∎

We want to note that $\nabla \eta_n(\theta) = O_P(1)$ and $\Delta_n = O_P(1/\sqrt{n})$.

Example 45 We shall assume that a random variable $X = 1, 0$ occurs with probabilities θ_* and $1 - \theta_*$. Let $0 < \epsilon < 0.5$ be known. If $\epsilon \le \theta_* \le 1 - \epsilon$ is unknown and we apply a statistical model with $\epsilon \le \theta \le 1 - \epsilon$, and if ones appear k times in n observations $x_1, \ldots, x_n$, then from

$$P(x|\theta) = \begin{cases} \theta, & x = 1 \\ 1 - \theta, & x = 0 \end{cases},$$

we obtain

$$\eta_n(\theta)/\sqrt{n} = -\theta_* \log \theta - (1 - \theta_*) \log(1 - \theta) + \frac{k}{n} \log \theta + (1 - \frac{k}{n}) \log(1 - \theta)$$

$$+ \theta_* \log \theta_* + (1 - \theta_*) \log(1 - \theta_*) - \frac{k}{n} \log \theta_* - (1 - \frac{k}{n}) \log(1 - \theta_*),$$

$$\nabla \eta_n(\theta)/\sqrt{n} = -\frac{\theta_*}{\theta} + \frac{1 - \theta_*}{1 - \theta} + \frac{k}{n\theta} - \frac{1}{1 - \theta}(1 - \frac{k}{n}) = \frac{1}{\theta(1 - \theta)} \frac{k}{n} - \frac{\theta_*}{\theta(1 - \theta)}$$

$$\nabla \eta_n(\theta_*)/\sqrt{n} = \frac{k/n - \theta_*}{\theta_*(1 - \theta_*)},$$

$$J(\theta) = \theta_* \frac{\partial^2}{\partial \theta^2}\{-\log \theta\} + (1 - \theta_*)\frac{\partial^2}{\partial \theta^2}\{-\log(1 - \theta)\}$$

$$= \theta_* \frac{1}{\theta^2} + (1 - \theta_*)\frac{1}{(1 - \theta)^2} = \frac{\theta_* - 2\theta\theta_* + \theta^2}{\theta^2(1 - \theta)^2},$$

and

$$I = J = J(\theta_*) = \frac{1}{\theta_*(1 - \theta_*)}.$$

Therefore, according to Proposition 10, as n increases,

$$\frac{\sqrt{n}(k/n - \theta_*)}{\theta_*(1 - \theta_*)} \xrightarrow{d} N(0, \frac{1}{\theta_*(1 - \theta_*)})$$

holds. This is equivalent to

$$\frac{\sqrt{n}(k/n - \theta_*)}{\sqrt{\theta_*(1 - \theta_*)}} \xrightarrow{d} N(0, 1)$$

and it can also be derived from the Central Limit Theorem. ∎

In the following, when the log-likelihood $f(\cdot, \theta) = \log \frac{p(\cdot|\theta_*)}{p(\cdot|\theta)}$ is an $L^2(q)$-valued analytic function, we call the function in Sect. 5.1

$$\eta_n(\theta) = \frac{1}{\sqrt{4}}\sum_{i=1}^{n}\{\mathbb{E}_X[f(X, \theta)] - f(x_i, \theta)\}$$

an empirical process, and consider its behavior as $n \to \infty$.

First, note that $\eta_1, \eta_2, \ldots$, given in (5.1), have the same mean and covariance. In fact, we have

$$\mathbb{E}_{X_1,\ldots,X_n}[\eta_n(\theta)] = 0, \quad \theta \in \Theta$$

$$\mathbb{E}_{X_1,\ldots,X_n}[\eta_n(\theta)\eta_n(\theta')] = \mathbb{E}_X[\log \frac{p(X|\theta_*)}{p(X|\theta)} \log \frac{p(X|\theta_*)}{p(X|\theta')}], \quad \theta, \theta' \in \Theta$$

which do not depend on n.

On the other hand, a random variable η that takes values in analytic functions is called a *Gaussian process* when $\eta(\theta_1), \ldots, \eta(\theta_m)$ follow a m-dimensional normal distribution for any positive integer m and any elements $\theta_1, \ldots, \theta_m$ of Θ. Especially when Θ is compact, it is known that a sequence of such random variables (empirical processes) η_n converges in distribution to a Gaussian process η.

Proposition 11 *Suppose the log-likelihood $f(\cdot, \theta)$ is an $L^2(q)$-valued analytic function. When Θ is compact, the empirical processes $\eta_1, \eta_2, \ldots$, converge in distribution to a Gaussian process η.*

Proposition 11 holds whether or not the true distribution q is regular with respect to the statistical model $p(\cdot|\theta)_{\theta\in\Theta}$. While the proof is omitted, we want to mention its significance. First, from the Central Limit Theorem, for any θ, there exists a constant σ^2 such that $\eta_n(\theta) \xrightarrow{d} N(0, \sigma^2)$. Also, the Central Limit Theorem can be extended to multiple dimensions. That is, for any $m \geq 1$ and $\theta_1, \ldots, \theta_m \in \Theta$, there exists a positive definite matrix $\Sigma \in \mathbb{R}^{m\times m}$ such that

$$(\eta_n(\theta_1), \ldots, \eta_n(\theta_m)) \xrightarrow{d} N(0, \Sigma). \tag{5.3}$$

However, it is not easy to prove that this dimension is extended to infinity and $\eta_n \xrightarrow{d} \eta$ holds. In fact, Proposition 11 implies the following proposition.

Proposition 12 Assume the log-likelihood $f(\cdot, \theta)$ is an analytic function taking values in $L^2(q)$. When Θ is compact, $\sup_{\theta\in\Theta}\{\eta_n(\theta)\}^2$ converges in distribution to $\sup_{\theta\in\Theta}\eta(\theta)^2$.

In fact, letting $C(\Theta)$ be the set of continuous functions in Θ, the function

$$h : C(\Theta) \ni \phi \mapsto \sup_{\theta\in\Theta}|\phi(\theta)| \in \mathbb{R}$$

is continuous (under the uniform norm):

$$|h(\phi_1) - h(\phi_2)| = |\sup_{\theta\in\Theta}|\phi_1(\theta)| - \sup_{\theta\in\Theta}|\phi_2(\theta)|| \leq \sup_{\theta\in\Theta}|\phi_1(\theta) - \phi_2(\theta)| = \|\phi_1 - \phi_2\| \tag{5.4}$$

(Problem 43). From this fact and the convergence in distribution $\eta_n \rightarrow \eta$ (Proposition 11), Proposition 12 holds (see page 151 of reference [12]). However, even if we prove (5.3), we cannot obtain Proposition 12. As Θ is an infinite set and the $\theta \in \Theta$ that reaches the supremum is different for each $n = 1, 2, \ldots$, even if each $\theta \in \Theta$ converges in distribution to a normal distribution, it does not necessarily converge to a Gaussian process as a whole.

5.2   Asymptotic Normality of the Posterior Distribution

In this section, we will show that in the regular case, as the sample size n increases, the posterior distribution follows a normal distribution.

In this section, we write $U_n \sim V_n$ to indicate the relationship between the sequences of random variables $\{U_n\}$ and $\{V_n\}$ whose ratio converge in probability to 1 as $n \rightarrow \infty$.[1] Also, we define the sequence ϵ_n as

$$\epsilon_n = n^{-1/4}. \tag{5.5}$$

---

[1] The symbol $\sim$ is often used to represent an equivalence relation between elements of a set S (Sect. 7.2). Outside of this section, it is also used in the sense of $X \sim N(0, 1)$ (the random variable X follows the standard normal distribution).

We assume that the true distribution is regular with respect to the statistical model, which means that Θ_* contains just one element. From now on, we define $B(\epsilon, \theta) := \{\theta' \in \Theta \mid \mathbb{E}_X\left[\log \frac{p(X|\theta)}{p(X|\theta')}\right] < \epsilon\}$, and partition Θ into two regions, $B(\epsilon_n, \theta)$ and its complement $B(\epsilon_n, \theta_*)^C$. Then, when we define

$$U_n^{(0)} := \int_{B(\epsilon_n, \theta_*)^C} \prod_{i=1}^{n} \frac{p(x_i|\theta)}{p(x_i|\theta_*)}\varphi(\theta)d\theta \text{ and } U_n^{(1)} := \int_{B(\epsilon_n, \theta_*)} \prod_{i=1}^{n} \frac{p(x_i|\theta)}{p(x_i|\theta_*)}\varphi(\theta)d\theta$$

we can ignore the value of the former (Proposition 13), and for the marginal likelihood

$$Z_n = \int_{\Theta} \prod_{i=1}^{n} p(x_i|\theta)\varphi(\theta)d\theta$$

we can show that $U_n^{(1)} \sim Z_n / \prod_{i=1}^{n} p(x_i|\theta_*)$ (Proposition 14).

Proposition 13 (Watanabe [10]) *When Θ is compact, the following two equations hold:*

$$U_n^{(0)} = o_P(\exp(-\sqrt{n})) \tag{5.6}$$

and

$$U_n^{(2)} := \int_{B(\epsilon_n, \theta_*)^C} \left\{\sum_{i=1}^{n} \log \frac{p(x_i|\theta_*)}{p(x_i|\theta)}\right\} \prod_{i=1}^{n} \frac{p(x_i|\theta)}{p(x_i|\theta_*)}\varphi(\theta)d\theta = o_P(\exp(-\sqrt{n})). \tag{5.7}$$

where Θ_* generally contains multiple elements (we do not assume regularity), and we arbitrarily fix one of them, θ_*.

Proof Refer to the appendix at the end of this chapter.

We will use Eq. (5.7) in Chap. 6.

Note that the proof of Proposition 13 does not use regularity. We will discuss the general case in Chap. 8, but only the $\theta \in \Theta$ included in $B(\epsilon_n, \theta_*)$ are subject to generalization.

Proposition 14 *When the true distribution is regular with respect to the statistical model,*

$$Z_n \sim \prod_{i=1}^{n} p(x_i|\theta_*)(\frac{2\pi}{n})^{d/2} \frac{\varphi(\theta_*)}{\sqrt{\det(J)}} \exp\left\{\frac{n}{2}\Delta_n^T J \Delta_n\right\} \tag{5.8}$$

is valid.

Proof Assuming that the true distribution is regular with respect to the statistical model, $\nabla\mathbb{E}_X[-\log p(X|\theta)] = 0$ holds. Also, for $\theta \in B(\epsilon_n, \theta_*)$, by Taylor's expansion around θ_*, there exist $\theta^1, \theta^2 \in B(\epsilon_n, \theta_*)$ such that

$$\mathbb{E}_X[-\log p(X|\theta)] = \mathbb{E}_X[-\log p(X|\theta_*)] + \frac{1}{2}(\theta - \theta_*)^\top J(\theta^1)(\theta - \theta_*) \quad (5.9)$$

and

$$\eta_n(\theta) = (\theta - \theta_*)^\top \nabla \eta_n(\theta^2)$$

hold. And, assuming that the true distribution is regular with respect to the statistical model, the matrix $J(\theta_*)$ is regular. Hence, since $\epsilon_n \to 0$, for sufficiently large n, we may assume that the matrix $J(\theta^1)$ is regular. That is,

$$\sum_{i=1}^{n} -\log p(x_i|\theta)$$

$$= n \left\{ \mathbb{E}_X[-\log p(X|\theta)] - \mathbb{E}_X[-\log p(X|\theta_*)] - \frac{1}{n}\sum_{i=1}^{n}\log p(x_i|\theta_*) - \frac{1}{\sqrt{n}}\eta_n(\theta) \right\}$$

$$= n \left\{ \mathbb{E}_X[-\log p(X|\theta_*)] + \frac{1}{2}(\theta - \theta_*)^\top J(\theta^1)(\theta - \theta_*) \right\}$$

$$- n \left\{ \mathbb{E}_X[-\log p(X|\theta_*)] + \frac{1}{n}\sum_{i=1}^{n}\log p(x_i|\theta_*) + \frac{1}{\sqrt{n}}(\theta - \theta_*)^\top \nabla \eta_n(\theta^2) \right\}$$

$$= \sum_{i=1}^{n} -\log p(x_i|\theta_*) + \frac{n}{2}(\theta - \theta_*)^\top J(\theta^1)(\theta - \theta_*) - \sqrt{n}(\theta - \theta_*)^\top \nabla \eta_n(\theta^2)$$

$$= \sum_{i=1}^{n} -\log p(x_i|\theta_*) + \frac{n}{2}\|J(\theta^1)^{1/2}\left\{\theta - \theta_* - \frac{J(\theta^1)^{-1}\nabla \eta_n(\theta^2)}{\sqrt{n}}\right\}\|^2 - \frac{n}{2}\|J(\theta^1)^{-1/2}\frac{\nabla \eta_n(\theta^2)}{\sqrt{n}}\|^2$$

can be established, where the final transformation employs the method of completing the square. This can be confirmed from the fact that

$$\|J(\theta^1)^{1/2}\{\theta - \theta_* - \frac{J(\theta^1)^{-1}\nabla \eta_n(\theta^2)}{\sqrt{n}}\}\|^2$$

$$= (\theta - \theta_*)^\top J(\theta^1)(\theta - \theta_*) - \frac{2}{\sqrt{n}}(\theta - \theta_*)^\top J(\theta^1)J(\theta^1)^{-1}\nabla \eta(\theta^2) + \|J(\theta^1)^{-1/2}\frac{\nabla \eta_n(\theta^2)}{\sqrt{n}}\|^2$$

holds. Also, as $n \to \infty$, $\theta^1, \theta^2 \to \theta_*$, so

$$\sum_{i=1}^{n} -\log p(x_i|\theta)$$

$$= \sum_{i=1}^{n} -\log p(x_i|\theta_*) + \frac{n}{2}\|J^{1/2}(\theta - \theta_* - \Delta_n)\|^2 - \frac{n}{2}\|J^{1/2}\Delta_n\|^2 + o_P(1)$$

$$= \sum_{i=1}^{n} -\log p(x_i|\theta_*) + \frac{n}{2}(\theta - \theta_* - \Delta_n)^\top J(\theta - \theta_* - \Delta_n) - \frac{n}{2}\Delta_n^\top J \Delta_n + o_P(1)$$

$$(5.10)$$

is valid. Moreover, the integrand of

$$U_n^{(1)} = \int_{B(\epsilon_n, \theta_*)} \exp\{\sum_{i=1}^n \log \frac{p(x_i|\theta)}{p(x_i|\theta_*)}\} \varphi(\theta) d\theta$$

$$\sim \int_{B(\epsilon_n, \theta_*)} \exp\left(-\frac{n}{2}(\theta - \theta_* - \Delta_n)^\top J(\theta - \theta_* - \Delta_n)\right) d\theta \cdot \exp\left\{\frac{n}{2}\Delta_n^\top J \Delta_n\right\} \varphi(\theta_*)$$

$$(5.11)$$

is proportional to the probability density function of a normal distribution with mean $\theta_* + \Delta_n$ and covariance matrix $(nJ)^{-1}$. And, since $\sqrt{n}\epsilon_n \to \infty$,

$$\mathbb{E}_X[\log \frac{p(X|\theta_*)}{p(X|\theta)}] = \frac{1}{2}(\theta - \theta_*)^\top J(\theta - \theta_*) + o(|\theta|^3) < \epsilon_n$$

includes the region where the integrand takes its main values. Therefore, as $n \to \infty$, the integral becomes the same value as the integral over $\mathbb{R}^d$.

To put it more concretely, as $n \to \infty$,

1. The integrand (probability density function of the normal distribution) converges to a function that takes values only at one point θ_* (with the speed of convergence to the mean θ_* being $O(n^{-1/2})$ and the standard deviation converging to 0 at a speed of $O(n^{-1/2})$).
2. The integration domain $B(\epsilon_n, \theta_*)$ converges to a single point θ_* (at a rate of $O(n^{-1/4})$).

Because 1. is faster than 2., as $n \to \infty$, the entire integration domain of the integrand is included in $B(\epsilon_n, \theta_*)$, and this integral value (due to the properties of the probability density function of the normal distribution) converges to 1. In fact, if 2. converges faster than 1., the integral of a function with an integral value of 1 in $\mathbb{R}^d$ would converge to a single point before the integration domain, so the integral value would not converge to 1 (it would converge to 0).

Therefore,

$$U_n^{(1)} \sim (\frac{2\pi}{n})^{d/2}(\det J)^{-1/2} \exp\left(\frac{n}{2}\Delta_n^\top J \Delta_n\right) \varphi(\theta_*)$$

which means from Proposition 13, we have $Z_n \sim U_n^{(1)}$, where the final transformation used

$$\frac{1}{(2\pi)^{d/2}(\det \Sigma)^{1/2}} \int_{\mathbb{R}^d} \exp\{-\frac{1}{2}(\theta - \mu)^\top \Sigma^{-1}(\theta - \mu)\} d\theta = 1.$$

That is, we set $\Sigma = (nJ)^{-1}$, $\mu = \theta_* + \Delta_n$, and $\det \Sigma = \frac{1}{n^d} \cdot \frac{1}{\det J}$. ∎

Note that in the proofs up to this point in this section, we are using the three conditions of regularity and the assumption of compactness of Θ.

1. There exists a unique $\theta_* \in \Theta$ such that $\Theta_* = \{\theta_*\}$.

2. The matrix $\left[\dfrac{\partial^2 D(q|p(\cdot|\theta))}{\partial \theta_i \partial \theta_j}|_{\theta=\theta_*} \right] \in \mathbb{R}^{d \times d}$ is positive definite.

3. There exists an open set $\tilde{\Theta}$ such that $\theta_* \in \tilde{\Theta} \subseteq \Theta$.

The corresponding parts are marked in blue. The assumption of compactness of Θ is used in Proposition 13.

From Proposition 14, we see that the posterior distribution $p(\theta|x_1, \ldots, x_n)$ asymptotically converges to

$$
p(\theta|x_1, \ldots, x_n) = \frac{\prod_{i=1}^n p(x_i|\theta)\varphi(\theta)}{\int_\Theta \prod_{i=1}^n p(x_i|\theta')\varphi(\theta')d\theta'}
$$

$$
\sim C \exp\left\{ -\frac{n}{2}(\theta - \theta_* - \Delta_n)^\top J (\theta - \theta_* - \Delta_n) \right\},
$$

where C is a constant. In other words, it converges to $N(\theta_* + \Delta_n, \frac{1}{n} J^{-1})$.

Example 46 In Example 45, $\Delta_n = k/n - \theta_*$ implies that the posterior distribution of θ converges to $N(\theta_* + \dfrac{k}{n} - \theta_*, \dfrac{\theta_*(1 - \theta_*)}{n})$.   ∎

Proposition 15 *When the true distribution is in a regular relationship with the statistical model, the posterior mean of the function $s : \Theta \to \mathbb{R}$ after obtaining the samples $x_1, \ldots, x_n$ is given by*

$$
\mathcal{E}[s(\theta)] := \int_\Theta s(\theta) p(\theta|x_1, \ldots, x_n) d\theta
$$

$$
\sim \frac{\int_\Theta s(\theta) \exp\left\{ -\frac{n}{2}(\theta - \theta_* - \Delta_n)^\top J (\theta - \theta_* - \Delta_n) \right\} d\theta}{\int_\Theta \exp\left\{ -\frac{n}{2}(\theta - \theta_* - \Delta_n)^\top J (\theta - \theta_* - \Delta_n) \right\} d\theta}. \tag{5.12}
$$

where Δ_n is defined by Eqs. (5.1) and (5.2), and we set $J := J(\theta_*)$.

From Proposition 15, the following holds. It plays an important role in the next section and in Chap. 5 as a consequence of the asymptotic normality of the posterior probability.

Proposition 16 *When the true distribution is in a regular relationship with the statistical model, the following five equations hold.[2]*

$$
\mathcal{E}[\theta] = \theta_* + \Delta_n + o_P(\frac{1}{\sqrt{n}}), \tag{5.13}
$$

$$
\mathcal{E}[(\theta - \theta_* - \Delta_n)^\top J (\theta - \theta_* - \Delta_n)] = \frac{d}{n} + o_P(\frac{1}{n}), \tag{5.14}
$$

---

[2] Even if we take the mean with X, there remains uncertainty due to $x_1, \ldots, x_n$, so it becomes a matter of dealing with random variables. This is also why we use things like $o_P(\cdot)$.

$$\mathcal{E}[(\theta - \theta_*)^\top J (\theta - \theta_*)] = \frac{d}{n} + \Delta_n^\top J \Delta_n + o_P(\frac{1}{n}), \tag{5.15}$$

$$\mathcal{E}(x) := \mathcal{E}[- \log p(x|\theta)] = - \log p(x|\theta_*) - \Delta_n^\top \nabla \log p(x|\theta_*) + o_P(\frac{1}{\sqrt{n}}), \tag{5.16}$$

and

$$\mathcal{V}(x) := \mathcal{E}[\{- \log p(x|\theta) - \mathcal{E}(x)\}^2]$$
$$= \frac{1}{n} \mathrm{tr} \left\{ J^{-1} \nabla(- \log p(x|\theta_*)) \nabla(- \log p(x|\theta_*))^\top \right\} + o_P(\frac{1}{n}). \tag{5.17}$$

Proof Refer to the appendix at the end of the chapter (apply Slutsky's theorem (Eqs. 4.8–4.10) and the Continuous Mapping Theorem (Eq. 4.11)).

Example 47 Standard normal random numbers were generated n times, and the posterior distributions of the mean μ and variance σ^2 were calculated, yielding shapes close to a normal distribution (Fig. 5.1a, b). Comparing the cases of $n = 100$ and $n = 10000$, it can be seen that the latter has vertical and horizontal shapes of the probability density function that are 10 times larger and 1/10 times smaller, respectively. The corresponding R code is as follows: ∎

```
1  library(rstan)
2  f <- function(N) stan("model6.stan", data = list(N = N, y = rnorm(N)))
3  # Chapter 2's model6.stan
4  fit1 <- f(100)        # n=100
5  fit2 <- f(10000)      # n=10000
6  stan_hist(fit1)
7  stan_hist(fit2)
```

Example 48 (*Mixture Normal Distribution*) We shall revisit the example of the mixture normal distribution from Example 14 in Chap. 1. The prior distribution was given as $\mu_1, \mu_2 \sim N(0, \sigma^2)$, $\mu_1 < \mu_2$, and the posterior distribution of μ_1, μ_2 was calculated after observing n samples. In this case, if $0 < a < 1$ is the mixture ratio, the probability density function becomes

$$a \frac{1}{\sqrt{2\pi}} \exp\left\{ -\frac{(x - \mu_1)^2}{2} \right\} + (1 - a) \frac{1}{\sqrt{2\pi}} \exp\left\{ -\frac{(x - \mu_2)^2}{2} \right\}.$$

In addition, the standard normal distribution was taken as the true distribution. It becomes a distribution that mixes the probability density functions of $N(\mu_1, \sigma)$ and $N(\mu_2, \sigma)$ at the ratio of a and $1 - a$. In this case, it does not become regular (Example 14). The results output by Stan showed that the posterior distribution deviated from the normal distribution (Fig. 5.1). The R code is as follows: ∎

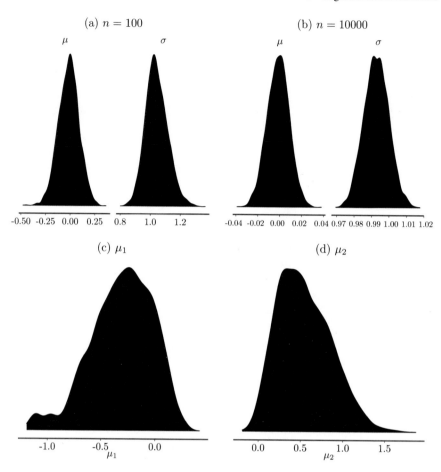

Fig. 5.1 In the case of regularity, the shape of the posterior distribution becomes a normal distribution. When comparing **a** $n = 100$ and **b** $n = 10000$, the variance in the latter case is 1/100 (the standard deviation is 1/10). **c, d** In the case where the true distribution is a normal distribution and the statistical model is a mixed normal distribution, it is not regular. It can be seen that the posterior distributions of the parameters μ_1, μ_2 deviate from the normal distribution

```
1  library(rstan)
2  N <- 100
3  y <- rnorm(100)
4  fit <- stan(file = "model13.stan", data = list(N=N, y=y))
5  stan_dens(fit, pars="mu")
```

5.3 Generalization Loss and Empirical Loss

The quantities

$$G_n := \mathbb{E}_X[-\log r(X|x_1, \ldots, x_n)]$$

and

$$T_n := \frac{1}{n} \sum_{i=1}^{n} \{-\log r(x_i|x_1, \ldots, x_n)\}$$

obtained from the predictive distribution $r(x|x_1, \ldots, x_n)$ are respectively called the *generalization loss* and *empirical loss*. These quantities play a crucial role in defining the WAIC in the next chapter.

Here, G_n is obtained by taking the average with respect to the random variable X, but like T_n, it depends on the samples $x_1, \ldots, x_n \in \mathcal{X}$. From this section onwards, we will consider the samples $x_1, \ldots, x_n$ as random variables (they should be written as $X_1, \ldots, X_n$), and regard G_n, T_n as random variables. However, we will use uppercase notation for $x_1, \ldots, x_n$ only when performing operations to take the mean or variance.

From Proposition 1, the inequality

$$\mathbb{E}_X[\log \frac{q(X)}{p(X|\theta)}] \geq 0 , \ \theta \in \Theta \tag{5.18}$$

holds between the true distribution q and the statistical model $\{p(\cdot|\theta)\}_{\theta \in \Theta}$. Statistics and learning theory correspond to the problem of constructing $\theta = \hat{\theta}(x_1, \ldots, x_n)$ from samples $x_1, \ldots, x_n \in \mathcal{X}$ to minimize the left-hand side of (5.18). And since $\mathbb{E}_X[\log q(X)]$ is a constant, the minimization of (5.18) is equivalent to the minimization of

$$\mathbb{E}_X[-\log p(X|\hat{\theta}(x_1, \ldots, x_n))].$$

At this time, it does not necessarily become the minimization, but it is often the case to take θ that maximizes

$$\sum_{i=1}^{n} \log p(x_i|\theta)$$

as $\hat{\theta}(x_1, \ldots, x_n)$ (maximum likelihood estimation). Moreover, in Watanabe's Bayesian theory, instead of constructing $\hat{\theta}(x_1, \ldots, x_n)$ from samples $x_1, \ldots, x_n \in \mathcal{X}$, it uses the predictive distribution r in Bayesian theory and sets

$$p(X|\hat{\theta}(x_1, \ldots, x_n)) = r(X|x_1, \ldots, x_n).$$

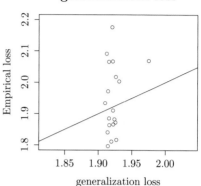

Fig. 5.2   Initially, in Example 49, we compared the exact values of generalization loss and empirical loss with the values obtained by Stan. As a result, it was found that the two values almost coincide (left). Also, we plotted the values of the generalization loss G_n and the empirical loss T_n. It was found that the variance of the generalization loss was smaller compared to the empirical loss (right)

The value obtained by taking $\mathbb{E}_X\left[-\log(\cdot)\right]$ of that value becomes the generalization loss. Also, because the true distribution q is unknown, it is sometimes approximated by

$$\frac{1}{n}\sum_{i=1}^{n}-\log[p(x_i|\hat{\theta}(x_1,\ldots,x_n))].$$

The latter is the empirical loss. Whether to use maximum likelihood estimation, generalization loss, or empirical loss will be discussed in the next chapter (Fig. 5.2).

Example 49   For Example 4, from (2.12), we have

$$-\log r(x|x_1,\ldots,x_n) = \frac{1}{2}\log(2\pi\frac{n+2}{n+1}) + \frac{n+1}{2(n+2)}\{x - \frac{1}{n+1}\sum_{i=1}^{n}x_i\}^2$$

and

$$\mathbb{E}_X[(X - \frac{1}{n+1}\sum_{i=1}^{n}x_i)^2] = 1 + (\mu - \frac{1}{n+1}\sum_{i=1}^{n}x_i)^2.$$

Then, G_n, T_n can be written as follows.

$$G_n = \frac{1}{2}\log(2\pi\frac{n+2}{n+1}) + \frac{n+1}{2(n+2)}\{1 + (\mu - \frac{1}{n+1}\sum_{i=1}^{n}x_i)^2\}$$

and

$$T_n = \frac{1}{2}\log(2\pi\frac{n+2}{n+1}) + \frac{n+1}{2(n+2)}\frac{1}{n}\sum_{i=1}^{n}\{x_i - \frac{1}{n+1}\sum_{j=1}^{n}x_j\}^2.$$

In the R language functions, they are as follows:

```
1  G <- function(y)
2     0.5*log(2*pi*(n+2)/(n+1)) + (n+1)/(n+2)/2*(sigma^2+(mu-sum(y)/(n+1))**
3     2)
```

```
1  T<- function(y)
2     0.5*log(2*pi*(n+2)/(n+1)) + (n+1)/(n+2)/2*mean((y-sum(y)/(n+1))**2)
```

∎

However, the ability to analytically solve the generalization loss G_n and empirical loss T_n, as in Example 49, is limited to cases such as using a conjugate prior distribution for the exponential family of distributions. Therefore, the same example is calculated using Stan. However, the generalization loss cannot be evaluated if the true distribution q is unknown, and the true distribution is assumed as

$$f(x) = \frac{1}{2}f(x|-1) + \frac{1}{2}f(x|1), \quad f(x|\mu) = \frac{1}{\sqrt{2\pi}}\exp\{-\frac{(x-\mu)^2}{2}\}.$$

Also, as mentioned in Chap. 2, the empirical loss T_n can generally be defined as follows.

```
1  T_n <- function(log_likelihood) -mean(log(colMeans(exp(log_likelihood))))
```

Each part has the following meaning.

exp(log_likelihood)	$p(x_i\|\theta)p(\theta\|x_1,\ldots,x_n), i = 1,\ldots,n$
colMeans(exp(log_likelihood))	Approximation of $\int_\Theta p(x_i\|\theta)p(\theta\|x_1,\ldots,x_n)d\theta$
Overall	$-\frac{1}{n}\sum_{i=1}^{n}\log p(x_i\|x_1,\ldots,x_n)$

Also, a function to find the generalization loss is as follows.

```
1  f_true <- function(x) 0.5*dnorm(x, -1, 1) + 0.5*dnorm(x, 1, 1)
```

```
1  G_n <- function(y_pred, f_true) {
2    dens <- density(y_pred)
3    f_pred <- approxfun(dens$x, dens$y, yleft=1e-15, yright=1e-15)
4    f_ge <- function(x) f_true(x)*(-log(f_pred(x)))
5    ge <- integrate(f_ge, lower=-6, upper=6)$value
6    return(ge)
7  }
```

Each part has the following meaning.

y_pred	m random numbers following the predictive distribution $r(x\|x_1,\ldots,x_n)$
dens	$(x^{(j)}, r(x^{(j)}\|x_1,\ldots,x_n)), j = 1,\ldots, m$
f_pred	Function $r(\cdot\|x_1,\ldots,x_n)$
f_true	Function $q(\cdot)$
f_ge	Function $-\log r(\cdot\|x_1,\ldots,x_n)q(\cdot)$
Overall	$\int_\mathcal{X} -\log r(x\|x_1,\ldots,x_n)q(x)dx$

And in the `generated quantities` block, various quantities can be computed. For instance, in the following `model14.stan`, not only the empirical loss, but also the generalization loss can be calculated. The Stan code was set as follows.

model14.stan

```
1   data{
2     int N; // Number of data
3     real y[N]; // Data
4   }
5
6   parameters{
7     real mu; // Mean
8   }
9
10  model{
11    mu ~ normal(0,100); // Prior distribution of mean
12    y ~ normal(mu, 1); // Vectorization
13  }
14
15  generated quantities {
16    vector[N] log_lik;
17    real y_pred;
18    for(n in 1:N)
19    log_lik[n] = normal_lpdf(y[n] | mu, 1);
20    y_pred = normal_rng(mu, 1);
21  }
```

And the following process was carried out.

```
1   n <- 100
2   ## Calculation of horizontal axis and vertical axis values
3   m <- 20
4   GG <- NULL
5   TT <- NULL
6   T_stan <- NULL
7   G_stan <- NULL
8   sigma <- 1
9   mu <- 0
10  for(j in 1:m){
11    y <- c(rnorm(n*0.5,-1,1),rnorm(n*0.5,1,1))
12    GG <- c(GG,G(y))
13    TT <- c(TT,T(y))
14    data_list <- list(N=n, y=y)
15    fit <- stan(file="model14.stan", data=data_list)
```

```
16    ms <- rstan::extract(fit)
17    T_stan <- c(T_stan,T_n(ms$log_lik))
18    G_stan <- c(G_stan,G_n(ms$y_pred, f_true))
19  }
20  ## Plotting the graph
21  plot(GG,G_stan,col="blue",xlim=c(1.8,2.2),ylim=c(1.8,2.2),xlab="Exact
22      values",ylab="Stan's values", main="Exact values and Stan's values")
23  abline(a=0,b=1)
24  points(TT,T_stan,col="red")
25  legend("topleft",legend=c("Generalization Loss","Empirical Loss"),pch=1,
26      col=c("blue","red"))
```

For a sample $x_1, \ldots, x_n$, fix $x \in \mathcal{X}$ and consider the function of $\alpha \in \mathbb{R}$:

$$s(x, \alpha) := \log \int_\Theta \left\{ \frac{p(x|\theta)}{p(x|\theta_*)} \right\}^\alpha p(\theta|x_1, \ldots, x_n)d\theta. \qquad (5.19)$$

Taylor expand this around $\alpha = 0$ to get

$$s(x, \alpha) = s(x, 0) + s'(x, 0)\alpha + \frac{1}{2}s''(x, 0)\alpha^2 + \sum_{k=3}^\infty \frac{1}{k!}s^{(k)}(x, 0)\alpha^k. \qquad (5.20)$$

Substitute $\alpha = 1$ to get

$$s(x, 1) = s(x, 0) + s'(x, 0) + \frac{1}{2}s''(x, 0) + \sum_{k=3}^\infty \frac{1}{k!}s^{(k)}(x, 0).$$

Accurately calculating both sides of this yields

$$-\log \int_\Theta p(x|\theta)p(\theta|x_1, \ldots, x_n)d\theta = \mathcal{E}(x) - \frac{1}{2}\mathcal{V}(x) - \sum_{k=3}^\infty \frac{1}{k!}s^{(k)}(x, 0), \quad (5.21)$$

where $\mathcal{E}(x)$ and $\mathcal{V}(x)$ are the mean and variance of $-\log p(x|\theta)$ due to the probabilistic variation of $\theta \in \Theta$ when sample $x_1, \ldots, x_n$ is obtained, and are quantities already defined in (5.16) and (5.17).

In fact, (5.21) is obtained by substituting $\alpha = 0$ into (5.20) and the following two equations:

$$s'(x, \alpha) = \frac{\int_\Theta \{\log p(x|\theta) - \log p(x|\theta_*)\}p(x|\theta)^\alpha p(\theta|x_1, \ldots, x_n)d\theta}{\int_\Theta p(x|\theta)^\alpha p(\theta|x_1, \ldots, x_n)d\theta} \qquad (5.22)$$

$$s''(x, \alpha) = \frac{\int_\Theta \{\log p(x|\theta) - \log p(x|\theta_*)\}^2 p(x|\theta)^\alpha p(\theta|x_1, \ldots, x_n) d\theta}{\int_\Theta p(x|\theta)^\alpha p(\theta|x_1, \ldots, x_n) d\theta}$$

$$- \frac{\{\int_\Theta \{\log p(x|\theta) - \log p(x|\theta_*)\} p(x|\theta)^\alpha p(\theta|x_1, \ldots, x_n) d\theta\}^2}{\{\int_\Theta p(x|\theta)^\alpha p(\theta|x_1, \ldots, x_n) d\theta\}^2} \tag{5.23}$$

(Exercise 51). Furthermore, we shall define

$$s_k(x, \alpha) := \frac{\mathcal{E}[\{\log p(x|\theta) - \log p(x|\theta_*)\}^k p(x|\theta)^\alpha]}{\mathcal{E}[p(x|\theta)^\alpha]}. \tag{5.24}$$

Under this definition, the following proposition holds true.

Proposition 17 For $s^{(k)}$, which is the kth derivative of the function $s(x, \alpha)$ with respect to α, there exists a positive constant C_k, for $k = 2, 3, \ldots$, such that

$$|s^{(k)}(x, \alpha)| \le C_k |s_k(x, \alpha)|. \tag{5.25}$$

Proof Refer to the appendix at the end of the chapter.

Utilizing this, the following proposition as well as Proposition 17 hold in general, without assuming regularity.

Proposition 18 *The generalization loss G_n and the empirical loss T_n can be expanded as follows:*

$$G_n = \mathbb{E}_X[\mathcal{E}(X)] - \frac{1}{2}\mathbb{E}_X[\mathcal{V}(X)] + o_P\left(\frac{1}{n}\right) \tag{5.26}$$

and

$$T_n = \frac{1}{n}\sum_{i=1}^n \mathcal{E}(x_i) - \frac{1}{2n}\sum_{i=1}^n \mathcal{V}(x_i) + o_P\left(\frac{1}{n}\right). \tag{5.27}$$

Proof By taking the expectation value of x as a random variable X in (5.21), we obtain

$$G_n = -\mathbb{E}[s(X, 1)] + \mathbb{E}_X[\log p(X|\theta_*)]$$

$$= \mathbb{E}_X[\mathcal{E}(X)] - \frac{1}{2}\mathbb{E}_X[\mathcal{V}(X)] - \mathbb{E}_X[\sum_{k=3}^\infty \frac{1}{k!} s^{(k)}(x, 0)]$$

From this, it suffices to show that the mean and the sample mean of $\frac{1}{k!}s^{(k)}(x, 0)$ in $\sum_{k=3}^\infty \frac{1}{k!} s^{(k)}(x, 0)$ are $O_P(n^{-k/2})$ which will be proved in Proposition 34.

Proposition 19 *When the true distribution is regular with respect to the statistical model, the generalization loss G_n and the empirical loss T_n can be expanded as follows:*

$$G_n = \mathbb{E}_X[-\log p(X|\theta)] + \frac{d}{2n} + \frac{1}{2}\Delta_n^\top J \Delta_n - \frac{1}{2n}\mathrm{tr}(IJ^{-1}) + o_P\left(\frac{1}{n}\right) \quad (5.28)$$

and

$$T_n = \frac{1}{n}\sum_{i=1}^{n} -\log p(x_i|\theta_*) + \frac{d}{2n} - \frac{1}{2}\Delta_n^\top J \Delta_n - \frac{1}{2n}\mathrm{tr}(IJ^{-1}) + o_P\left(\frac{1}{n}\right). \quad (5.29)$$

Proof Since (5.16) is accurate to $o_P(1/\sqrt{n})$, we use a method that does not use (5.16) to determine $\mathbb{E}_X[\mathcal{E}(X)]$ and $\frac{1}{n}\sum_{i=1}^{n}\mathcal{E}(x_i)$ to $o_P(1/n)$. First, by the Taylor expansion around θ_* and the fact that $\theta - \theta_* = o_P(1)$, we have

$$-\log p(x|\theta) = -\log p(x|\theta) + (\theta - \theta_*)^\top \nabla - \log p(x|\theta_*)$$
$$+ \frac{1}{2}(\theta - \theta_*)^\top \nabla^2 - \log p(x|\theta_*)(\theta - \theta_*) + o(1), \quad x \in \mathcal{X}$$

and

$$\mathbb{E}_X[-\log p(X|\theta)] = \mathbb{E}X[-\log p(X|\theta_*)] + \frac{1}{2}(\theta - \theta_*)^\top J(\theta_*)(\theta - \theta_*) + o(\|\theta - \theta_*\|^2),$$

where $o(\|\theta - theta_*\|^2)$ is a function $g(\theta)$ such that $g(\theta)/\|\theta - theta_*\|^2$ converges to zero as $\theta \to \theta_*$. Then, from (5.15),

$$\mathbb{E}_X[\mathcal{E}(X)] = \mathbb{E}_X[-\log p(X|\theta)] + \frac{d}{2n} + \frac{1}{2}\Delta_n^\top J \Delta_n + o\left(\frac{1}{n}\right)$$

is established. Similarly, from (5.10) and (5.14),

$$\frac{1}{n}\sum_{i=1}^{n}\mathcal{E}(x_i) = \frac{1}{n}\sum_{i=1}^{n} -\log p(x_i|\theta_*) + \frac{d}{2n} - \frac{1}{2}\Delta_n^\top J \Delta_n + o_P\left(\frac{1}{n}\right)$$

holds. Furthermore, from (5.17) and the definition of the matrix $I := I(\theta_*)$, we have

$$\mathbb{E}_X[\mathcal{V}(X)] = \frac{1}{n}\mathrm{tr}(IJ^{-1}) + o_P\left(\frac{1}{n}\right).$$

Also, applying the law of large numbers (Proposition 6) to

$$\mathcal{V}(x_i) = \frac{1}{n}\mathrm{tr}[\nabla(-\log p(x_i|\theta_*))\nabla(-\log p(x_i|\theta_*))^\top J^{-1}],$$

we obtain

$$\frac{1}{n}\sum_{i=1}^{n}\frac{\partial \log p(x_i|\theta_j)}{\partial \theta_j}\cdot\frac{\partial \log p(x_i|\theta_k)}{\partial \theta_k}\xrightarrow{P}I(\theta)j,k$$

for $j,k = 1,\ldots,d$, and from (4.8), we have

$$\frac{1}{n}\sum_{i=1}^{n}\mathcal{V}(x_i)=\frac{1}{n}\sum_{i=1}^{n}\frac{1}{n}\text{tr}(I+o_P(1))J^{-1}+o_P(\frac{1}{n})\frac{1}{n}\text{tr}(IJ^{-1})+o_P(\frac{1}{n}).$$

This completes the proof.   ∎

Appendix: Proof of Proposition

Proof of Proposition 13[3]

From the inequality of arithmetic and geometric means, we have

$$\sqrt{n}\eta_n(\theta)\leq\frac{1}{2}\left(n\epsilon_n+\frac{\sup_{\theta\in\Theta}\eta_n(\theta)^2}{\epsilon_n}\right).$$

Moreover, from the definition of η_n, for $\theta\in B(\epsilon_n,\theta_*)^C$, we obtain

$$\frac{1}{n}\sum_{i=1}^{n}\log\frac{p(x_i|\theta_*)}{p(x_i|\theta)}=\mathbb{E}_X\left[\log\frac{p(X|\theta_*)}{p(X|\theta)}\right]-\frac{\eta_n(\theta)}{\sqrt{n}}\geq\epsilon_n-\frac{\eta_n(\theta)}{\sqrt{n}}.$$

Therefore,

$$U_n^{(0)}\leq\int_{B(\epsilon_n,\theta_*)^C}\exp(-n\epsilon_n+\sqrt{n}\eta_n(\theta))\varphi(\theta)d\theta$$

$$\leq\exp(-n\epsilon_n+\sqrt{n}\sup_{\theta\in\Theta}\eta_n(\theta))\int_{B(\epsilon_n,\theta_*)^C}\varphi(\theta)d\theta\leq\exp\left(-\frac{n\epsilon_n}{2}+\frac{\sup_{\theta\in\Theta}\eta_n^2(\theta)}{2\epsilon_n}\right)$$

holds. Also, since Θ is compact (Assumption 3), according to Proposition 12, $\sup_{\theta\in\Theta}\eta_n^2(\theta)$ converges in law to a certain random variable. And, since $\epsilon_n=n^{-1/4}$, the exponent part becomes

$$-\frac{n\epsilon_n}{2}+\frac{\sup_{\theta\in\Theta}\eta_n^2(\theta)}{2\epsilon_n}=-\frac{1}{2}n^{3/4}+\frac{1}{2}n^{1/4}\cdot O_P(1)=-\frac{1}{2}n^{3/4}+o_P(n^{3/4}),$$

---

[3] The proof of Lemma 1 in [10] was referred to.

where we used $\sup_{\theta \in \Theta} \eta_n(\theta) = O_P(1)$, i.e., $\sup_{\theta \in \Theta} \eta_n^2(\theta) = O_P(1)$. Therefore, (5.6) holds.

Furthermore, since Θ is compact, $M := \sup_{\theta \in \Theta} \mathbb{E}_X[\log \frac{p(X|\theta_*)}{p(X|\theta)}]$ is bounded, and $\|\eta_n\| := \sup_{\theta \in \Theta} |\eta_n(\theta)|$ converges in law to a certain random variable. Therefore, we can write

$$|\sum_{i=1}^n \log \frac{p(x_i|\theta_*)}{p(x_i|\theta)}| = |n\mathbb{E}_X[\log \frac{p(X|\theta_*)}{p(X|\theta)}] - \sqrt{n}\eta_n(\theta)| \leq n(M + \|\eta_n\|/\sqrt{n}).$$

Moreover, applying (4.9), we have

$$U_n^{(2)} \leq n\{M + \|\eta_n\|/\sqrt{n}\}U_n^{(0)} = O_P(1) \cdot \exp(\log n) \cdot o_P(\exp(-\sqrt{n})) = o_P(\exp(-\sqrt{n})).$$

∎

Proof of Proposition 16

1. Since $\theta \sim N(\theta_* + \Delta_n, (nJ)^{-1})$ and $\Delta_n = O_P(1/\sqrt{n})$, we have $\mathcal{E}[\theta - \theta_*] = \Delta_n + o_P(1/\sqrt{n})$. Since $\sqrt{n}\mathcal{E}[\theta - \theta_*] - \Delta_n$ converges in probability to 0, $\mathcal{E}[\theta] - \theta_* - \Delta_n = o_P(1/\sqrt{n})$ holds.
2. From (4.10), we have $o_P(1/\sqrt{n})o_P(1/\sqrt{n}) = o_P(1/n)$. Moreover, using 1., we can get

$$\mathcal{E}[(\theta - \theta_* - \Delta_n)^\top J(\theta - \theta_* - \Delta_n)]$$
$$= \mathcal{E}[\text{tr}\{J(\theta - \theta_* - \Delta_n)(\theta - \theta_* - \Delta_n)^\top\}]$$
$$= \text{tr}\{J\mathcal{E}[(\theta - \mathcal{E}[\theta] + o_P(1/\sqrt{n}))(\theta - \mathcal{E}[\theta] + o_P(1/\sqrt{n}))^\top]\}$$
$$= \text{tr}\{J\mathcal{E}[(\theta - \mathcal{E}[\theta])(\theta - \mathcal{E}[\theta])^\top] + \text{tr}(J \cdot o_P(\frac{1}{n})I)\}$$
$$= \text{tr}\{J(nJ)^{-1}\} + o_P(\frac{1}{n}) = \frac{d}{n} + o_P(\frac{1}{n}), \quad (5.30)$$

where we used the fact that for matrices $A \in \mathbb{R}^{m \times n}$, $B \in \mathbb{R}^{n \times m}$, the traces of their products AB and BA are equal (Exercise 53). Also, since $o_P(1/\sqrt{n})$ does not contain terms related to θ, we have

$$\mathcal{E}[o_P(1/\sqrt{n})\theta - \mathcal{E}[\theta]] = o_P(1/\sqrt{n})\mathcal{E}[\theta - \mathcal{E}[\theta]] = 0.$$

Note that while $\mathbb{E}[s(\theta)]$ denotes the posterior mean of $s : \Theta \to \mathbb{R}$. However, in the case of $s : \Theta \to \mathbb{R}^{d \times d}$, it will mean the posterior mean for each component.
3. This can be derived by expanding (5.30).
4. From the mean value theorem, there exists some $\theta^1 \in \Theta$ between θ and θ_* such that

$$\log p(x|\theta) = \log p(x|\theta_*) + (\theta - \theta_*)^\top \nabla \log p(x|\theta^1)$$

and from (4.11), as $n \to \infty$, the posterior mean of $(\theta - \theta_*)^\top \nabla \log p(x|\theta^1)$ approaches $\mathcal{E}[\theta - \theta_*]^\top \nabla \log p(x|\theta_*)$. Therefore, the posterior mean of the second term on the right-hand side also approaches the same value. Hence,

$$
\begin{aligned}
\mathcal{E}[\log p(x|\theta)] &= \mathcal{E}[\{\log p(x|\theta_*) + (\theta - \theta_*)^\top \nabla \log p(x|\theta_*)(1 + o_P(1))\}] \\
&= \log p(x|\theta_*) + \mathcal{E}[\theta - \theta_*]^\top \nabla \log p(x|\theta_*)(1 + o_P(1)) \\
&= \log p(x|\theta_*) + \{\Delta_n + o_P(\frac{1}{\sqrt{n}})\}^\top \nabla \log p(x|\theta_*)(1 + o_P(1)) \\
&= \log p(x|\theta_*) + \Delta_n^\top \nabla \log p(x|\theta_*) + o_P(\frac{1}{\sqrt{n}}).
\end{aligned}
$$

Thus, (5.16) can be derived. Note that we have applied $o_P(1/\sqrt{n})o_P(1) = o_P(1/\sqrt{n})$ which can be derived from (4.10).

5. First, from

$$\log p(x|\theta) = \log p(x|\theta_*) + (\theta - \theta_*)^\top (\nabla \log p(x|\theta_*) + o(1)),$$

we obtain

$$
\begin{aligned}
\mathcal{V}[\log p(x|\theta) - \log p(x|\theta_*)] &= \mathcal{V}[(\theta - \theta_*)^\top (\nabla \log p(x|\theta_*) + o_P(1))] \\
&= \mathcal{E}[\{(\theta - \theta_*)^\top \nabla \log p(x|\theta_*) + o_P(1)\}^2] - \mathcal{E}[\log p(x|\theta) - \log p(x|\theta_*)]^2.
\end{aligned}
$$
$$(5.31)$$

where $\mathcal{V}[\cdot]$ represents the operation of posterior variance with respect to θ. Moreover, taking the posterior mean of

$$
\begin{aligned}
&\{(\theta - \theta_*)^\top (\nabla \log p(x|\theta_*) + o_P(1))\}^2 \\
&= \mathrm{tr}(\theta - \theta_*)(\theta - \theta_*)^\top \nabla \log p(x|\theta_,)\nabla \log p(x|\theta_*)^\top (1 + o_P(1))
\end{aligned}
$$

from (5.15), we obtain

$$
\begin{aligned}
&\mathcal{E}[\{(\theta - \theta_*)^\top (\nabla \log p(x|\theta_*) + o_P(1))\}^2] \\
&= \mathrm{tr}(\frac{J^{-1}}{n} + \Delta_n \Delta_n^\top)\nabla \log p(x|\theta_*)\nabla \log p(x|\theta_*)^\top + o_P(\frac{1}{n}).
\end{aligned}
$$
$$(5.32)$$

Furthermore, similar to the derivation of (5.16), we have

$$\mathcal{E}[\log p(x|\theta) - \log p(x|\theta_*)] = \Delta_n^\top \nabla \log p(x|\theta_*) + o_P(\frac{1}{\sqrt{n}}). \qquad (5.33)$$

Finally, from (5.31), subtracting the square of (5.33) from (5.32), we obtain (5.17).

∎

Proof of Proposition 17

Generally, each term of $s^{(k)}(x, \alpha)$ is the product $\prod_h s_h(x, \alpha)$ in (5.24), such that $s'(x, \alpha) = s_1(x, \alpha)$, $s''(x, \alpha) = s_2(x, \alpha) - s_1(x, \alpha)^2$, etc., where the sum of the degrees of each term equals k. Indeed, if this holds for some $k \geq 1$, then

$$
\begin{aligned}
&s'h(x, \alpha)\\
&= \frac{\int_\Theta \{\log \frac{p(x|\theta)}{p(x|\theta_*)}\}^{h+1} p(x|\theta)^\alpha p(\theta|x_1, \ldots, x_n) d\theta}{\int_\Theta p(x|\theta)^\alpha p(\theta|x_1, \ldots, x_n) d\theta}\\
&\quad - \frac{\{\int_\Theta \{\log \frac{p(x|\theta)}{p(x|\theta_*)}\}^h p(x|\theta)^\alpha p(\theta|x_1, \ldots, x_n) d\theta\}\{\int_\Theta \{\log \frac{p(x|\theta)}{p(x|\theta_*)}\} p(x|\theta)^\alpha p(\theta|x_1, \ldots, x_n) d\theta\}}{\{\int_\Theta p(x|\theta)^\alpha p(\theta|x_1, \ldots, x_n) d\theta\}^2}\\
&= s_{h+1}(x, \alpha) - s_h(x, \alpha) s_1(x, \alpha).
\end{aligned}
$$

That is, each term of $s^{(k+1)}(x, \alpha)$ also has a sum of degrees equal to $k + 1$, and it takes the form

$$s^{(3)}(x, \alpha) = s_3(x, \alpha) - 3s_2(x, \alpha)s_1(x, \alpha) + 2s_1(x, \alpha)^3. \tag{5.34}$$

Therefore, each term of $s^{(k)}(x, \alpha)$ is a product of $s_h(x, \alpha)$.
By Hölder's inequality,

$$|s_i(x, \alpha)|^{1/i} \leq |s_{i+j}(x, \alpha)|^{1/(i+j)}$$

we have[4] $|s_{i+j}(x, \alpha)| \geq |s_i(x, \alpha)s_j(x, \alpha)|$, $i, j = 1, 2, \ldots$. Therefore, $|s_k(x, \alpha)|$ is the maximum in the product $\prod_h s_h(x, \alpha)$ excluding the coefficients of each term of $s^{(k)}(x, \alpha)$, and (5.25) holds. Here, C_k is the sum of the absolute values of the coefficients of each term of $s^{(k)}(x, \alpha)$, and it holds that $C_2 = 2, C_3 = 6, \ldots$. For example, the coefficients of $s^{(3)}(x, \alpha)$ in (5.34) are $1, -3, 2$, so $C_3 = |1| + |-3| + |2| = 6$. ∎

Exercises 42–53

42. Let the random variable $X = 1, 0$ occur with probabilities $\theta_*, 1 - \theta_*$. Suppose $0 < \theta_* < 1$ is unknown and we apply a statistical model with $0 < \theta < 1$, demonstrate the following:

$$I = J = J(\theta_*) = \frac{1}{\theta_*(1 - \theta_*)}.$$

43. Show the following inequalities to prove (5.4).

---

[4] This can be seen as a generalization of the fact that the square mean of a random variable is never less than the square of its mean (since the variance is nonnegative).

(a) $|\phi_1(\theta)| \le |\phi_2(\theta)| + |\phi_1(\theta) - \phi_2(\theta)|$

(b) $\sup_{\theta \in \Theta} |\phi_1(\theta)| \le \sup_{\theta \in \Theta} |\phi_2(\theta)| + \sup_{\theta \in \Theta} |\phi_1(\theta) - \phi_2(\theta)|$.

44. Confirm the following three inequalities used in the proof of Proposition 13.

$$U_n^{(0)} \le \int_{B(\epsilon_n, \theta_*)^C} \exp(-n\epsilon_n + \sqrt{n}\eta_n(\theta))\varphi(\theta)d\theta$$

$$\le \exp(-n\epsilon_n + \sqrt{n} \sup_{\theta \in \Theta} \eta_n(\theta)) \int_{B(\epsilon_n, \theta_*)^C} \varphi(\theta)d\theta \le \exp\left(-\frac{n\epsilon_n}{2} + \frac{\sup_{\theta \in \Theta} \eta_n^2(\theta)}{2\epsilon_n}\right).$$

45. Confirm the following equalities used to derive (5.8). That is, for $\theta \in B(\epsilon_n, \theta_*)$, $\theta_* \in \Theta_*$, we have

$$\mathbb{E}_X[-\log p(X|\theta)] = \mathbb{E}_X[-\log p(X|\theta_*)] + \frac{1}{2}(\theta - \theta_*)^\top J(\theta^1)(\theta - \theta_*)$$

$$\eta_n(\theta) = (\theta - \theta_*)^\top \nabla \eta_n(\theta^2),$$

where $\theta^1, \theta^2 \in B(\epsilon_n, \theta_*)$ exist and condition

$$\|J(\theta^1)^{1/2}\{\theta - \theta_* - \frac{J(\theta^1)^{-1}\nabla\eta_n(\theta^2)}{\sqrt{n}}\}\|^2$$

$$= (\theta - \theta_*)^\top J(\theta^1)(\theta - \theta_*) - \frac{2}{\sqrt{n}}(\theta - \theta_*)^\top J(\theta^1)J(\theta^1)^{-1}\nabla\eta(\theta^2) + \|J(\theta^1)^{-1/2}\frac{\nabla\eta_n(\theta^2)}{\sqrt{n}}\|^2$$

is satisfied.

46. In the process of obtaining Proposition 14, where are the three conditions of regularity applied?

(a) There exists (uniquely) $\theta_* \in \Theta$ such that $\Theta_* = \{\theta_*\}$

(b) The matrix $\left[\dfrac{\partial^2 D(q|p(\cdot|\theta))}{\partial\theta_i\partial\theta_j}\right]_{\theta=\theta_*} \in \mathbb{R}^{d\times d}$ is positive definite

(c) There exists an open set $\tilde{\Theta}$ such that $\theta_* \in \tilde{\Theta} \subseteq \Theta$

47. In Fig. 5.1, when comparing (a) $n = 100$ and (b) $n = 10000$, why is the variance in the latter 1/100 (standard deviation 1/10)? Also, why are the posterior distributions of parameters μ_1, μ_2 in (c)(d) deviating from the normal distribution?

48. In Example 4, show that the generalization loss and empirical loss are respectively given by

$$G_n = \frac{1}{2}\log(2\pi\frac{n+2}{n+1}) + \frac{n+1}{2(n+2)}\{1 + (\mu - \frac{1}{n+1}\sum_{i=1}^{n}x_i)^2\}$$

and

$$T_n = \frac{1}{2}\log(2\pi\frac{n+2}{n+1}) + \frac{n+1}{2(n+2)}\frac{1}{n}\sum_{i=1}^{n}\{x_i - \frac{1}{n+1}\sum_{j=1}^{n}x_j\}^2.$$

Also, write the code that generates G_n and T_n, generate 100 standard normal random numbers with $\mu = 0$, and calculate these values.

49. Why can the generalization loss be obtained from the y_pred in the generated quantities block via the function below? Explain the role of each step after dens.

```
1  G_n <- function(y_pred, f_true) {
2    dens <- density(y_pred)
3    f_pred <- approxfun(dens$x, dens$y, yleft=1e-15, yright=1e-15)
4    f_ge <- function(x) f_true(x)*(-log(f_pred(x)))
5    ge <- integrate(f_ge, lower=-6, upper=6)$value
6    return(ge)
7  }
```

50. Prove Eq. (5.21) from (5.20), (5.22) and (5.23).
51. Show that each term of $s^{(k)}(x, \alpha)$, which is the kth derivative of (5.19) with respect to α, can be written as a product of $s_h(x, \alpha), h \leq k$, defined in (5.24).
52. Explain how to apply (4.8) in the last step of the proof of Proposition 19.
53. For matrices $A \in \mathbb{R}^{m \times n}$ and $B \in \mathbb{R}^{n \times m}$, prove that the traces of their products AB and BA are equal.

Chapter 6
Information Criteria

In this chapter, we discuss information criteria such as AIC and BIC. In Watanabe's Bayesian theory, new information criteria, such as WAIC and WBIC, are proposed. Existing information criteria assume that the true distribution is regular with respect to the statistical model, and they cannot be applied to general situations. In this chapter, we point out that this is due to their definition using maximum likelihood estimates. Indeed, although WAIC is defined using empirical loss, it does not use maximum likelihood estimates. Furthermore, we clarify that AIC, TIC, and WAIC show almost the same performance when assuming regularity. In addition, we introduce WBIC, which corresponds to a generalization of BIC, and clarify that it shows similar performance when assuming regularity. Note that there is the following relationship among the propositions presented in this chapter.

Relationship between the propositions of Chapter 6 and regularity

Regular	→	P.10	→	P.21	→	P.24	←	P.19	←	Regular
↓	↘			↓					↙	
P.14		P.20	→	P.22	→	P.23	←	P.9		
↓	↖									
P.26	←	P.13								

(Red text is applicable only in Chaps. 5 and 6, blue text is applicable in Chaps. 8 and 9 as well).

6.1 Model Selection Based on Information Criteria

Suppose there are statistical models Θ, Θ', and data $x_1, \ldots, x_n \in \mathcal{X}$ have been observed. We want to identify which model the observations came from. Here, we define the smallest d for which $\Theta \subseteq \mathbb{R}^d$ as the *dimension of the parameter space*, denoted as $d(\Theta)$. When the dimensions of the parameter spaces $d(\Theta)$ and $d(\Theta')$

© The Author(s), under exclusive license to Springer Nature Singapore Pte Ltd. 2023   121
J. Suzuki, *WAIC and WBIC with R Stan*,
https://doi.org/10.1007/978-981-99-3838-4_6

are different, it is difficult to determine which of Θ or Θ' is more appropriate based solely on the *likelihood*. In particular, if there is an inclusion relationship such as $\Theta \subseteq \Theta'$, the goodness of fit of Θ' will be better, and its likelihood will be larger than that of Θ. In fact, the following holds:

$$\Theta \subseteq \Theta' \implies \sup_{\theta \in \Theta} \prod_{i=1}^{n} p(x_i|\theta) \leq \sup_{\theta \in \Theta'} \prod_{i=1}^{n} p(x_i|\theta).$$

However, Θ' is more complex as a model. In other words, the following holds:

$$\Theta \subseteq \Theta' \implies d(\Theta) \leq d(\Theta').$$

We should evaluate both the goodness of fit of the data $x_1, \ldots, x_n \in \mathcal{X}$ to Θ and Θ', and the complexity of Θ and Θ'. In this chapter, we call the sum of the quantities representing the goodness of fit of the data to the statistical model and the complexity of the statistical model the *information criterion*, and we examine the problem of selecting the statistical model that minimizes this value.

In the following, we call the θ that maximizes the likelihood $\prod_{i=1}^{n} p(x_i|\theta)$, or minimizes the negative log-likelihood

$$l := -\frac{1}{n} \sum_{i=1}^{n} \log p(x_i|\theta)$$

the maximum likelihood estimation.[1]

In this chapter, among the information criteria we discuss, *AIC* (Akaike Information Criterion), *BIC* (Bayesian Information Criterion), and *TIC* (Takeuchi Information Criterion) use the quantity

$$\frac{1}{n} \sum_{i=1}^{n} \{-\log p(x_i|\hat{\theta})\}$$

to represent the goodness of fit to the statistical model. As for the complexity of the statistical model, AIC applies $d(\Theta)$, BIC applies $\frac{d(\Theta)}{2} \log n$, and TIC applies $\mathrm{tr}(I_n J_n^{-1})$. Here, we define

$$J_n = \left(-\frac{1}{n} \sum_{k=1}^{n} \frac{\partial^2 \log p(x_k|\theta)}{\partial \theta_i \partial \theta_j} \right)_{i,j} \bigg|_{\theta=\hat{\theta}} \quad \text{and} \quad I_n = \frac{1}{n} \sum_{k=1}^{n} \nabla \log p(x_k|\hat{\theta}) \{\nabla \log p(x_k|\hat{\theta})\}^{\top}.$$

In other words, we define $d := d(\Theta)$ and use the quantities

---

[1] If the maximum likelihood does not exist, neither do the maximum likelihood estimation nor the values of the information criterion that can be represented using it.

$$AIC := \frac{1}{n} \sum_{i=1}^{n} \{-\log p(x_i|\hat{\theta})\} + \frac{d}{n}, \tag{6.1}$$

$$BIC := \frac{1}{n} \sum_{i=1}^{n} \{-\log p(x_i|\hat{\theta})\} + \frac{d}{2n} \log n, \tag{6.2}$$

and

$$TIC := \frac{1}{n} \sum_{i=1}^{n} \{-\log p(x_i|\hat{\theta})\} + \frac{1}{n} \text{tr}(I_n J_n^{-1}) \tag{6.3}$$

to compare the statistical models Θ and Θ' in each of the information criteria, where $\hat{\theta}$ is the maximum likelihood estimator of θ based on $x_1, \ldots, x_n \in \mathcal{X}$.

In this book, in addition to that, we also consider information criteria such as the *free energy*, *WAIC* (Watanabe Akaike Information Criterion), and *WBIC* (Watanabe Bayesian Information Criterion).[2] In particular, we aim to clarify the principles on which these criteria are based. In this section, as preparation for that, we will look at examples of model selection using AIC and BIC.

Example 50 (Multiple Regression) Suppose we are given observed data (x_1, y_1), $\ldots, (x_n, y_n) \in \mathbb{R}^p \times \mathbb{R}$.[3] We aim to find $\beta_0 \in \mathbb{R}$ and $\beta \in \mathbb{R}^p$ that minimize

$$\sum_{i=1}^{n} (y_i - \beta_0 - \sum_{j=1}^{p} x_{i,j} \beta_j)^2.$$

In practice, we perform centering by subtracting $\frac{1}{n} \sum_{k=1}^{n} x_{k,j}$ from each $x_{i,j}$ and $\frac{1}{n} \sum_{k=1}^{n} y_k$ from each y_i, and set the intercept β_0 to 0. Using the obtained $\hat{\beta} = [\hat{\beta}_1, \ldots, \hat{\beta}_p]^\top$ and the newly calculated $\hat{\beta}_0 := y_i - \sum_{j=1}^{p} x_{i,j} \hat{\beta}_j$, we predict the dependent variable y for a new set of covariates observation $x = [x^{(1)}, \ldots, x^{(p)}]^\top$ as $y = \hat{\beta}_0 + \sum_{j=1}^{p} x^{(j)} \hat{\beta}_j$.

During this process, we consider the likelihood of the underlying statistical model. Instead of using all p covariates, we aim to minimize the values of AIC and BIC by selecting a subset of d variables ($0 \le d \le p$). It is important to note that although the likelihood is maximized when $d = p$, minimizing the values of AIC and BIC does not necessarily correspond to this maximum likelihood. This is because AIC and BIC take into account the model's complexity in addition to the goodness of fit to the data, such as likelihood. Here, when minimizing

$$l = \frac{1}{n} \sum_{i=1}^{n} -\log p(x_i, y_i|\beta, \sigma^2) = \frac{1}{2} \log 2\pi\sigma^2 + \frac{1}{n} \sum_{i=1}^{n} \frac{(y_i - \sum_{k=1}^{p} x_{i,k} \beta_k)^2}{2\sigma^2}$$

---

[2] Professor Sumio Watanabe himself refers to them as Widely Applicable Information Criterion and Widely Applicable Bayesian Information Criterion, respectively [13].

[3] For multiple regression, refer to "Statistical Learning with Math and R" by Suzuki Joe (Springer, 2020) [14].

obtained from

$$p(x_i, y_i | \beta, \sigma^2) = \frac{1}{\sqrt{2\pi\sigma^2}} \exp\{-\frac{(y_i - \sum_{k=1}^{p} x_{i,k}\beta_k)^2}{2\sigma^2}\},$$

differentiating as

$$\frac{\partial l}{\partial \beta_j} = -\frac{1}{n} \sum_{i=1}^{n} \frac{x_{i,j}(y_i - \sum_{k=1}^{p} x_{i,k}\beta_k)}{\sigma^2}$$

and

$$\frac{\partial l}{\partial \sigma^2} = \frac{1}{2\sigma^2} - \frac{1}{n} \sum_{i=1}^{n} \frac{(y_i - \sum_{k=1}^{p} x_{i,k}\beta_k)^2}{2(\sigma^2)^2},$$

we find that the maximum likelihood estimators are $\hat{\beta}_1, \dots, \hat{\beta}_p$ and

$$\hat{\sigma}^2 = \frac{1}{n} \sum_{i=1}^{n} (y_i - \sum_{k=1}^{p} x_{i,k}\hat{\beta}_k)^2 ,$$

which minimize $\sum_{i=1}^{n} (y_i - \sum_{k=1}^{p} x_{i,k}\beta_k)^2$.
In that case, the value of l is given by

$$\frac{1}{2} \log 2\pi\hat{\sigma}^2 + \frac{1}{2}. \tag{6.4}$$

And then, considering the value of AIC, by adding $\frac{p}{n}$ to this and replacing p with $1 \le d \le p$, it becomes the minimization of

$$\frac{1}{2} \log \hat{\sigma}^2 + \frac{d}{n} + \frac{1}{2} \log 2\pi + \frac{1}{2}. \tag{6.5}$$

Since $\frac{1}{2} \log 2\pi$ and $\frac{1}{2}$ are constants, multiplying by $2n$ results in minimization[4] of

$$n \log \hat{\sigma}^2 + 2d \tag{6.6}$$

Similarly, BIC can be reduced to the minimization of

$$n \log \hat{\sigma}^2 + d \log n.$$

---

[4] In cases where equivalent minimization can be obtained by adding constants or multiplying by constants from the original AIC value, the converted value is often called AIC as well. The same applies to BIC.

The code for calculating and plotting AIC and BIC is shown below. There are 2^p subsets of $\{1, \ldots, p\}$, and for each, the maximum likelihood estimates $(\hat{\beta}, \hat{\sigma}^2)$ are determined. The function that finds the minimum sum of squares for each d is RSS.min, and the function that calculates the information criterion such as AIC and BIC is IC. ∎

```
1   RSS.min <- function(X, y, T) {
2     m <- ncol(T)
3     S.min <- Inf
4     for (j in 1:m) {
5       q <- T[, j]
6       S <- sum((lm(y ~ X[, q])$fitted.values - y) ^ 2)
7       if (S < S.min) {
8         S.min <- S
9         set.q <- q
10      }
11    }
12    return(list(value = S.min, set = set.q))
13  }
```

```
1   IC <- function(k) {
2     T <- combn(1:p, k)  # The rows of matrix T are each one of the size k
                subsets of {1,...,p}.
3     res <- RSS.min(X, y, T)
4     AIC <- n * log(res$value / n) + 2 * k
5     BIC <- n * log(res$value / n) + k * log(n)
6     return(list(AIC = AIC, BIC = BIC))
7   }
```

Out of the 13 explanatory variables in the CRAN package Boston (average housing price dataset), discrete ones are excluded from the beginning, and $p = 11$ is set.

```
1   library(MASS)
2   df <- Boston
3   X <- as.matrix(df[, c(1, 3, 5, 6, 7, 8, 10, 11, 12, 13)])
4   y <- df[[14]]
5   n <- nrow(X)
6   p <- ncol(X)
7   AIC.seq <- NULL
8   BIC.seq <- NULL
9   for (k in 1:p) {
10    AIC.seq <- c(AIC.seq, IC(k)$AIC)
11    BIC.seq <- c(BIC.seq, IC(k)$BIC)
12  }
13  plot(1:p, ylim = c(min(AIC.seq), max(BIC.seq)), type = "n", xlab = "# of
14       variables", ylab = "IC values")
15  lines(AIC.seq, col = "red")
16  lines(BIC.seq, col = "blue")
17  legend("topright", legend = c("AIC", "BIC"), col = c("red", "blue"), lwd
18       = 1, cex = .8)
19  which(AIC.seq==min(AIC.seq))
20  which(BIC.seq==min(BIC.seq))
```

The results are shown in Fig. 6.1.

Fig. 6.1 AIC and BIC
values (vertical axis) for the
Boston dataset (Table 3.1).
The horizontal axis is the
number of selected variables.
Both AIC and BIC draw
downward convex curves. In
particular, the point where
AIC is minimized is further
to the right

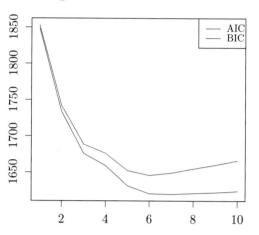

6.2   AIC and TIC

To understand AIC and TIC, it is necessary to clarify the behavior of the average log-likelihood. First, we shall look at how the estimates obtained by maximum likelihood approach θ_*.

Proposition 20 *Assume that the true distribution has a regular relationship with the statistical model. When Θ is compact, the maximum likelihood estimate $\hat{\theta}_n$ converges in probability to the true parameter θ_* as $n \to \infty$, and*

$$\hat{\theta}_n = \theta_* + \Delta_n + o_P\left(\frac{1}{\sqrt{n}}\right).$$

Proof See the appendix at the end of the chapter.

Proposition 20 was derived to guarantee the performance of AIC and TIC in the regular case (Propositions 22 and 23). Therefore, it does not apply to the performance guarantee of WAIC or the subsequent chapters.

In the following, we will only consider the maximum likelihood estimate $\hat{\theta}$, and in order to emphasize that it varies with the samples $x_1, \ldots, x_n \in \mathcal{X}$, we will write it as $\hat{\theta}_n$ or $\hat{\theta}(x_1, \ldots, x_n)$. Furthermore, for Δ_n and J_n, when we explicitly specify the samples, we will write them as $\Delta(x_1, \ldots, x_n)$ and $J(x_1, \ldots, x_n)$, respectively. In this chapter, we will apply the following proposition several times.

Proposition 21 *Assume that the true distribution has a regular relationship with the statistical model. In this case, when $x_1, \ldots, x_n \in \mathcal{X}$ are considered as random variables $X_1, \ldots, X_n$ following the distribution q, and their averages are taken,*

$$n\mathbb{E}_{X_1,\ldots,X_n}[\Delta(X_1 \ldots X_n)^\top J \Delta(X_1 \ldots X_n)] = \text{tr}(IJ^{-1}) + o(1) \qquad (6.7)$$

holds.

Proof From the definition of Δ_n (5.2),

$$n\Delta_n^\top J \Delta_n = \nabla\eta_n(\theta_*)^\top J^{-1}\nabla\eta_n(\theta_*) = \text{tr}[\nabla\eta_n(\theta_*)\nabla\eta_n(\theta_*)^\top J^{-1}]$$

holds (Exercise 53). Furthermore, Proposition 10 implies that $\mathbb{E}_{X_1-X_n}[\nabla\eta_n(\theta_*) \nabla\eta_n(\theta_*)^\top] \to I$. Therefore, (5.7) holds. ∎

Example 51 Consider binary logistic regression. For the covariate $x \in \mathbb{R}^p$, the probability of occurrence of $y \in \{1, -1\}$ is expressed using the intercept $\beta_0 \in \mathbb{R}$ and the slope $\beta \in \mathbb{R}^p$ as

$$P(y|x) = \frac{1}{1 + \exp -y(\beta_0 + x\beta)}.$$

We want to find $\beta_0 \in \mathbb{R}$ and $\beta \in \mathbb{R}^p$ that maximize the likelihood $\prod_{i=1}^n P(y_i|x_i)$ from the training data $(x_1, y_1), \ldots, (x_n, y_n) \in \mathbb{R}^p \times \{1, -1\}$. In this case, if $\Theta \subseteq \mathbb{R}^{p+1}$ containing $\theta := \begin{bmatrix} \beta_0 \\ \beta \end{bmatrix}$ is not compact, there is a possibility that the θ that maximizes the likelihood does not exist. For example, if for all $i = 1, \ldots, n$, $y_i(\beta_0 + x_i\beta) > 0$, then by doubling β_0 and β, the value of $y_i(\beta_0 + x_i\beta)$ is doubled. That is, for all i, $P(y_i|x_i)$ increases, and the likelihood can be approached to 1 as much as desired. This occurs when the sample size n is small, and in the case where the maximum likelihood estimate does not exist, the consistency argument does not apply. In the case where the optimal (β_0, β) is infinite, it is not regular (there exists an open set $\tilde{\Theta} \subseteq \Theta$ containing θ_*). Conversely, if Θ is restricted to a compact set, a $\hat{\theta}_n$ that maximizes the likelihood can be found even in such cases. ∎

Example 52 Example 45 corresponds to the regular case. Therefore, for $\Delta_n = k/n - \theta_*$, the maximum likelihood estimator $k/n = \theta_* + \Delta_n$ converges in probability to θ_*. ∎

AIC and TIC can be calculated from the samples $x_1, \ldots, x_n \in \mathcal{X}$, and are designed to be unbiased estimators of a certain quantity when taking the average of test data and training data (respectively, $\mathbb{E}_X[\cdot]$, $\mathbb{E}_{X_1 \ldots X_n}[\cdot]$). In particular, the expectation over the training data is an operation of the average related to the maximum likelihood estimation (it can also be written as $\mathbb{E}_{\hat{\theta}(X_1 \ldots X_n)}[\cdot]$ instead of $\mathbb{E}_{X_1 \ldots X_n}[\cdot]$). Proposition 20, which describes the behavior of maximum likelihood estimation under the assumption of regularity, plays an essential role in deriving the following proposition.

Proposition 22 *When the true distribution is regular with respect to the statistical model,*

$$\mathbb{E}_X[-\log p(X|\hat{\theta}(x_1, \ldots, x_n))]$$
$$= \mathbb{E}_X[-\log p(X|\theta_*)] + \frac{1}{2}\Delta(x_1, \ldots, x_n)^\top J \Delta(x_1, \ldots, x_n) + o_P(\frac{1}{n}) \quad (6.8)$$

and

$$\frac{1}{n}\sum_{i=1}^n \{-\log p(x_i|\hat{\theta}(x_1, \ldots, x_n))\}$$
$$= \frac{1}{n}\sum_{i=1}^n \{-\log p(x_i|\theta_*)\} - \frac{1}{2}\Delta(x_1, \ldots, x_n)^\top J \Delta(x_1, \ldots, x_n) + o_P(\frac{1}{n}) \quad (6.9)$$

hold. Furthermore, when $X, X_1, \ldots, X_n$ *are independent random variables following the distribution* q,[5] *the following hold, respectively:*

$$U := \mathbb{E}_{X_1 \ldots X_n}\mathbb{E}_X[-\log p(X|\hat{\theta}(X_1, \ldots, X_n))]$$
$$= \mathbb{E}_X[-\log p(X|\theta_*)] + \frac{1}{2n}\text{tr}(IJ^{-1}) + o(\frac{1}{n}) \quad (6.10)$$

and

$$\mathbb{E}_{X_1 \ldots X_n}[\frac{1}{n}\sum_{i=1}^n \{-\log p(X_i|\hat{\theta}(X_1, \ldots, X_n))\}]$$
$$= \mathbb{E}_X[-\log p(X|\theta_*)] - \frac{1}{2n}\text{tr}(IJ^{-1}) + o(\frac{1}{n}). \quad (6.11)$$

Proof Using (6.7), (6.8) (6.9) imply (6.10) (6.11) respectively. Therefore, it is sufficient to show (6.8) (6.9). From (4.18), Proposition 20, and the Mean Value Theorem, exists θ^1 on the line segment between $\hat{\theta}_n$ and θ such that

---

[5] In (6.8) (6.9), $X_1, \ldots, X_n$ are treated as random variables, so the error $o_P(\frac{1}{n})$ is allowed. However, in (6.10) (6.11), since the average is taken with respect to $X_1, \ldots, X_n$, there is no such random fluctuation, and the error becomes $o(\frac{1}{n})$.

$$\mathbb{E}_X[-\log p(X|\hat{\theta}_n)]$$
$$= \mathbb{E}_X[-\log p(X|\theta_*)] + (\hat{\theta}_n - \theta_*)^\top \nabla \mathbb{E}_X[-\log p(X|\theta_*)]$$
$$+ \frac{1}{2}(\hat{\theta}_n - \theta_*)^\top \nabla^2 \mathbb{E}_X[-\log p(X|\theta^1)](\hat{\theta}_n - \theta_*)$$
$$= \mathbb{E}_X[-\log p(X|\theta_*)] + \frac{1}{2}\Delta_n^\top J \Delta_n + o_P(\frac{1}{n}),$$

where the last equality is due to

$$\hat{\theta}_n \xrightarrow{P} \theta_* \Longrightarrow \theta^1 \xrightarrow{P} \theta_*$$

and (4.10) (4.11). Furthermore, from Proposition 20 and the Mean Value Theorem, and from (5.13), there exists θ^2 on the line segment between $\hat{\theta}_n$ and θ_* such that

$$\frac{1}{n}\sum_{i=1}^n \{-\log p(x_i|\theta_*)\} = \frac{1}{n}\sum_{i=1}^n \{-\log p(x_i|\hat{\theta}_n)\} + (\theta_* - \hat{\theta}_n)^\top \nabla[\frac{1}{n}\sum_{i=1}^n \{-\log p(x_i|\hat{\theta}_n)\}]$$
$$+ \frac{1}{2}(\theta_* - \hat{\theta}_n)^\top \nabla^2 \left[\frac{1}{n}\sum_{i=1}^n \{-\log p(x_i|\theta^2)\}\right](\theta_* - \hat{\theta}_n)$$
$$= \frac{1}{n}\sum_{i=1}^n \{-\log p(x_i|\hat{\theta}_n)\} + \frac{1}{2}\Delta_n^\top J \Delta_n + o_P(\frac{1}{n})$$

which holds true, where the last equality is obtained by applying (4.9) and the Continuous Mapping Theorem (4.11):

$$\hat{\theta}_n \xrightarrow{P} \theta_* \Longrightarrow \theta^2 \xrightarrow{P} \theta_* \Longrightarrow \nabla^2 \frac{1}{n}\sum_{i=1}^n \{-\log p(x_i|\theta^2)\} \xrightarrow{P} \nabla^2 \frac{1}{n}\sum_{i=1}^n \{-\log p(x_i|\theta_*)\}$$

while the Weak Law of Large Numbers (Proposition 6) is also used.

∎

Calculate the maximum likelihood estimator $\hat{\theta}(x_1, \ldots, x_n)$ for $x_1, \ldots,$ $x_n \in \mathcal{X}$, and evaluate it with the log-likelihood for $x_1', \ldots, x_n' \in \mathcal{X}$, taking the average:

$$\mathbb{E}_{X_1 \ldots X_n} \mathbb{E}_{X_1' \ldots X_n'} [\frac{1}{n}\sum_{i=1}^n -\log p(X_i'|\hat{\theta}(X_1, \ldots, X_n))].$$

which is (6.10). Calculate the maximum likelihood estimator $\hat{\theta}(x_1, \ldots, x_n)$ for $x_1, \ldots, x_n \in \mathcal{X}$, and evaluate it with the log-likelihood for the same $x_1, \ldots, x_n$, taking the average, which is (6.11). To give an analogy in stock price prediction, evaluating the predictive performance using the stock prices up to yesterday learned from the stock prices up to yesterday (6.11) is different from evaluating the learning results using the stock prices for tomorrow (6.10), and the former is better evaluated. Indeed, (6.9) is smaller than (6.8) by $\Delta(x_1, \ldots, x_n)^\top J \Delta(x_1, \ldots, x_n)$, and (6.11) is smaller than (6.10) by $\frac{1}{n} \mathrm{tr}(I J^{-1})$.

In AIC and TIC, the aim is to minimize the value of U in (6.10) by adding $\frac{1}{n} \mathrm{tr}(I J^{-1})$ to $\frac{1}{n} \sum_{i=1}^{n} - \log p(x_i | \hat{\theta}_n)$ in order to compensate for this difference. In other words, the justification for AIC and TIC is that their average coincides with (6.10).

Proposition 23 *Assume that the true distribution is regular with respect to the statistical model.*

$$\mathbb{E}_{X_1 \ldots X_n}[TIC] = U + o(\frac{1}{n}). \tag{6.12}$$

Furthermore, if realizable,

$$\mathbb{E}_{X_1 \ldots X_n}[AIC] = U + o(\frac{1}{n}) \tag{6.13}$$

holds true.

Proof From (6.10) (6.11), we have

$$\mathbb{E}_{X_1 \ldots X_n} \mathbb{E}_X[-\log p(X | \hat{\theta}_n(X_1, \ldots, X_n))]$$

$$= \mathbb{E}_{X_1 \ldots X_n}[\frac{1}{n} \sum_{i=1}^{n} \{-\log p(X_i | \hat{\theta}_n(X_1, \ldots, X_n))\}] + \frac{1}{n} \mathrm{tr}(I J^{-1}) + o(\frac{1}{n}).$$

$$\tag{6.14}$$

On the other hand, the second term of TIC is

$$I_n J_n^{-1} = I J^{-1} + o_P(1)$$

and we have

$$\mathbb{E}_{X_1 \ldots X_n}[I_n J_n^{-1}] = I J^{-1} + o(1)$$

which means (6.12). From Proposition 9, when realizable, $I = J$, so (6.13) is a special case. ∎

Example 53 Assuming that the true distribution q is realizable with respect to the statistical model $p(\cdot | \theta) \theta \in \Theta$ and can be written as

$$p(x|\theta) = \frac{1}{\sqrt{2\pi\sigma^2}} \exp\left\{-\frac{(x-\theta)^2}{2\sigma^2}\right\}, \ \theta \in \Theta$$

then we have

$$\sum_{i=1}^{n}(x_i - \theta)^2 = \sum_{i=1}^{n}(x_i - \theta_*)^2 - 2(\theta - \theta_*)\sum_{i=1}^{n}(x_i - \theta_*) + n(\theta - \theta_*)^2$$

and

$$-\sum_{i=1}^{n}\log p(x_i|\theta) = \frac{n}{2}\log(2\pi\sigma^2) + \frac{1}{2\sigma^2}\sum_{i=1}^{n}(x_i - \theta)^2$$

$$= \frac{n}{2}\log(2\pi\sigma^2) + \frac{1}{2\sigma^2}\sum_{i=1}^{n}(x_i - \theta_*)^2 - \frac{1}{\sigma^2}(\theta - \theta_*)\sum_{i=1}^{n}(x_i - \theta_*) + \frac{n}{2\sigma^2}(\theta - \theta_*)^2$$

Assuming that $x_1, \ldots, x_n$ are realizations of independent random variables X generated by q, taking the average yields

$$n\mathbb{E}_X[-\log p(X|\theta)] = \frac{n}{2}\log(2\pi\sigma^2) + \frac{n}{2} + \frac{n}{2\sigma^2}(\theta - \theta_*)^2.$$

Substituting the maximum likelihood estimator $\hat{\theta}(x_1, \ldots, x_n) = \frac{1}{n}\sum_{i=1}^{n} x_i$ for θ, we get

$$\mathbb{E}_X[-\log p(X|\hat{\theta}(x_1, \ldots, x_n))] = \frac{1}{2}\log(2\pi\sigma^2) + \frac{1}{2} + \frac{1}{2\sigma^2}(\hat{\theta}(x_1, \ldots, x_n) - \theta_*)^2.$$

Assuming that $x_1, \ldots, x_n$ are realizations of independent random variables $X_1, \ldots, X_n$ generated by q, taking the average yields

$$U = \mathbb{E}_{X_1 \ldots X_n}\mathbb{E}_X[-\log p(X|\hat{\theta}(X_1, \ldots, X_n))] = \frac{1}{2}\log(2\pi\sigma^2 e) + \frac{1}{2n}. \quad (6.15)$$

This value corresponds to the AIC when $d = 1$. However, the value of σ^2 is unknown, and by calculating

$$\hat{\sigma}^2(x_1, \ldots, x_n) := \frac{1}{n}\sum_{i=1}^{n}(x_i - \hat{\theta}_n)^2$$

and

$$V := \log \sigma^2 - \mathbb{E}_{X_1 \ldots X_n}[\log \hat{\sigma}^2(X_1, \ldots, X_n)],$$

we obtain

$$AIC := \frac{1}{2} \log(2\pi e) + \frac{1}{2} \log \hat{\sigma}^2(x_1, \ldots, x_n) + \frac{V}{2} + \frac{1}{2n}. \qquad (6.16)$$

where $V = 2/n + o(1/n)$.

In fact, $\dfrac{n\hat{\sigma}^2(X_1, \ldots, X_n)}{\sigma^2}$ follows the χ^2 distribution with $n - 1$ degrees of freedom,[6] with mean and variance being $n - 1$ and $2(n - 1)$, respectively (Exercise
6.4). Therefore, using the Taylor expansion of $\log x$, $\log x = \log 1 + (x - 1)\frac{1}{1} - \frac{1}{2}(x - 1)^2 \frac{1}{1} + \ldots$, we can take the mean of

$$\log \frac{n\hat{\sigma}^2}{(n-1)\sigma^2} = 0 + \left(\frac{n\hat{\sigma}^2}{(n-1)\sigma^2} - 1 \right) - \frac{1}{2} \cdot \frac{1}{(n-1)^2} \left\{ \frac{n\hat{\sigma}^2}{\sigma^2} - (n-1) \right\}^2 + \ldots$$

and obtain

$$\mathbb{E}_{X_1 - X_n} [\log \frac{n\hat{\sigma}^2}{\sigma^2}] = \log(n - 1) - 0 - \frac{1}{2} \frac{1}{(n-1)^2} \cdot 2(n-1) + \cdots = \log(n - 1) - \frac{1}{n-1} + O(\frac{1}{n^2}).$$

This means that

$$nV = n\mathbb{E}_{X_1 \ldots X_n} [\log n - \log \frac{n\hat{\sigma}_n^2}{\sigma^2}]$$

$$= n \log n - n\{\log(n - 1) - \frac{1}{n-1} + O(\frac{1}{n^2})\} = 2 + o(1) \qquad (6.17)$$

using the equation

$$\log(n - 1) = \log n - \frac{1}{n} + O(\frac{1}{n^2}).$$

Thus, excluding $o(1/n)$, we have

$$AIC = \frac{1}{2} \log(2\pi e) + \frac{1}{2} \log \hat{\sigma}^2(x_1, \ldots, x_n) + \frac{3}{2n} \qquad (6.18)$$

■

Here, the χ^2 distribution with m degrees of freedom is the distribution of the sum of
squares of m independent standard normal random variables $X_1, \ldots, X_m \sim N(0, 1)$,
i.e., $\sum_{i=1}^{m} X_i^2$. The shape of the distribution is shown in Fig. 6.2. The following R
code was used to generate the figure:

---

[6] This is a well-known theorem in statistics, and proofs can be found in sources such as [16].

χ^2distribution

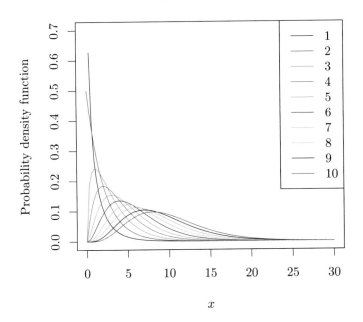

Fig. 6.2 Probability density functions of χ^2 distributions with degrees of freedom $1 \sim 10$

```
1  plot(0,xlim=c(0,30),ylim=c(0,0.7),xlab="$x$",ylab="Probability density
      function",
2    type="n", main="$\chi^2$ distribution")
3  for(i in 1:10)curve(dchisq(x,i),col=i, add=TRUE)
4  legend("topright",legend=1:10,col=1:10,lwd=1)
```

Furthermore, from Example 2, the square mean of each of $X_1, \ldots, X_m$ is 1, and the fourth moment is 3, so the mean of each of $X_1^2, \ldots, X_m^2$ is 1, and the variance is 2. Therefore, the mean of $X_1^2 + \cdots + X_m^2$ is m, and the variance is $2m$.

The analysis in Example 53 corresponds to the special case of multiple regression with p variables, where the number of variables is $p = 0$ (only an intercept, with no slope). In the general case of $p \neq 0$, we have

$$U = \frac{1}{2} \log(2\pi\sigma^2 e) + \frac{p+1}{2n}$$

$$AIC = \frac{1}{2} \log(2\pi\hat{\sigma}^2 e) + \frac{2p+3}{2n}$$

as shown.[7] When using AIC for variable selection, the minimization becomes $0 \leq d \leq p$

$$AIC(d) = \frac{1}{2}\log(2\pi\hat{\sigma}^2 e) + \frac{2d+3}{2n}$$

which is equivalent to minimizing (6.6). The purpose of calculating the exact value of (6.18) is to compare it with the value of WAIC in the next section.

6.3 WAIC

Before defining WAIC, we would like to mention a fact that justifies its existence. Watanabe's Bayesian theory claims that by replacing the empirical loss $\frac{1}{n}\sum_{i=1}^{n} -\log p(x_i|\theta)$ with T_n and replacing U with the average of the generalized loss G_n with respect to $X_1 \ldots X_n$, a similar relationship as AIC or TIC holds:

$$\mathbb{E}_{X_1\ldots X_n}[G_n] = \mathbb{E}_{X_1\ldots X_n}[T_n] + \frac{1}{n}\mathrm{tr}(IJ^{-1}) + o(\frac{1}{n}). \tag{6.19}$$

In fact, by setting $\lambda := \frac{d}{2}$ and $\nu := \frac{1}{2}\mathrm{tr}(IJ^{-1})$, Proposition 19 and Proposition 21 lead to the following proposition:

Proposition 24 *When the true distribution is regular with respect to the statistical model,*

$$\mathbb{E}_{X_1\ldots X_n}[G_n] = \mathbb{E}_X[-\log p(X|\theta_*)] + \frac{\lambda}{n} + o(\frac{1}{n})$$

and

$$\mathbb{E}_{X_1\ldots X_n}[T_n] = \mathbb{E}_X[-\log p(X|\theta_*)] + \frac{\lambda - 2\nu}{n} + o(\frac{1}{n})$$

hold.

Therefore, from (6.10) and (6.11), the following also hold:

$$\mathbb{E}_{X_1\ldots X_n}[G_n] = \mathbb{E}_{X_1\ldots X_n}\mathbb{E}_X[-\log p(X|\hat{\theta}(X_1,\ldots,X_n))] + \frac{\lambda - \nu}{n} + o(\frac{1}{n})$$

and

$$\mathbb{E}_{X_1\ldots X_n}[T_n] = \mathbb{E}_{X_1\ldots X_n}[\frac{1}{n}\sum_{i=1}^{n}\{-\log p(X_i|\hat{\theta}(X_1,\ldots,X_n))\}] + \frac{\lambda - \nu}{n} + o(\frac{1}{n}).$$

---

[7] This is proven in Chap. 5 of *Statistical Learning with Math and R*, Springer by Joe Suzuki.

Next, WAIC is defined as

$$WAIC_n := T_n + \frac{1}{n}\sum_{i=1}^{n}\mathcal{V}(x_i). \qquad (6.20)$$

And, we have

$$\mathbb{E}_{X_1\ldots X_n}\left[\frac{1}{n}\sum_{i=1}^{n}\mathcal{V}(X_i)\right] = \mathbb{E}_X[\mathcal{V}(X)]$$

$$= \frac{1}{n}\text{tr}\left\{J^{-1}\mathbb{E}_X[\nabla\log p(X|\theta_*)\nabla\log p(X|\theta_*)^\top]\right\} + o(\frac{1}{n})$$

$$= \frac{1}{n}\text{tr}(J^{-1}I) + o(\frac{1}{n}) \qquad (6.21)$$

in Proposition 38. The definition in (6.20) can be used similarly even when the true distribution is not regular with respect to the statistical model.

Therefore, when the true distribution is regular with respect to the statistical model, AIC, TIC, and WAIC coincide in terms of the first and second terms, excluding $O(1/n)$. Here, Proposition 23 and Proposition 24 guarantee the similar performance of AIC, TIC, and WAIC, but there is an essential difference. The former values are calculated based on maximum likelihood estimation, so the proof of Proposition 20 is necessary. The latter does not use maximum likelihood estimation, so by generalizing Proposition 19 alone, the performance in non-regular cases is also guaranteed. It is no exaggeration to say that the secret of WAIC's success lies in breaking away from the constraints of maximum likelihood estimation. In fact, even if it is not regular, WAIC can show the following generalized relationship of (6.19):

Proposition 25

$$\mathbb{E}_{X_1\ldots X_n}[G_n] = \mathbb{E}_{X_1\ldots X_n}[WAIC_n] + o(\frac{1}{n}).$$

Proposition 25 will be proved in Chap. 8.

The variance $V_n := \frac{1}{n}\sum_{i=1}^{n}\mathcal{V}(x_i)$ can be described as follows:

```
V_n <- function(log_likelihood):
    mean(colMeans(log_likelihood^2) - colMeans(log_likelihood)^2)
```

Using the previously defined T_n and the variance above, WAIC can be written in R language as follows:

```
WAIC <- function(log_likelihood) T_n (log_likelihood) + V_n (log_likelihood)
```

We want to verify that the values of AIC and WAIC coincide, excluding $O_P(1/n)$, in the regular case. To guarantee regularity, the true distribution q is required, so we will use artificial data.

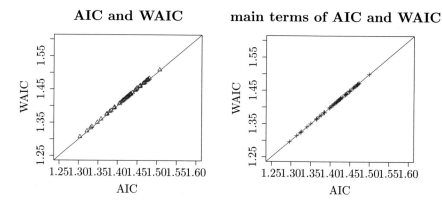

Fig. 6.3   The values of AIC and WAIC are almost identical. Also, the values of the term excluding $\frac{3}{2n}$ in AIC and T_n are almost identical

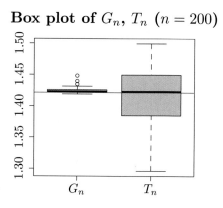

Fig. 6.4   The variance of T_n is larger than that of G_n. The red horizontal line represents the value of U

Example 54   Assuming the true distribution q follows $N(0, 1)$ and the statistical model is $N(\mu, \sigma^2)$, AIC, U, WAIC, and G_n are calculated 50 times. The R code is written as follows. The comparison of the overall values of AIC and WAIC, as well as the comparison of the first terms, is shown in Fig. 6.4. In this case, (6.18) was used for the AIC value. Overall, WAIC is slightly larger (Fig. 6.3 left), but the first term was in complete agreement (Fig. 6.3 right). The arithmetic mean values of AIC and WAIC for $m = 50$ times are obtained as follows, where the value of U, 1.421439, was calculated from (6.15).

U	Mean of AIC	Mean of WAIC	Mean of G_n
1.421439	1.422835	1.425567	1.424936

Furthermore, we found that the variance of G_n is very small. Theoretically, only the coincidence of the averages of both is guaranteed in the regular case, but it is often the case that AIC and WAIC are almost identical, as in this example.   ∎

Using `model14.stan` defined in Chap. 5, the following processing was performed.

```
model14 <- stan_model("model14.stan")
f_true <- function(x) dnorm(x)
T_stan <- NULL; G_stan <- NULL; V_stan <- NULL; AIC=NULL
m <- 50
n <- 200
for(j in 1:m){
  y <- rnorm(n,0,1)
  value <- 1/2*log (2*pi*exp(1))+1/2*log(var(y)/n*(n-1))+3/n/2
  AIC <- c(AIC,value)
  data_list <- list(N=n, y=y)
  fit <- sampling(model14, data=data_list, iter=6000, warmup=1000, chain
     =1,seed=123)
  ms <- rstan::extract(fit)
  T_stan <- c(T_stan,T_n(ms$log_lik))
  V_stan <- c(V_stan,V_n(ms$log_lik))
  G_stan <- c(G_stan,G_n(ms$y_pred,f_true))
}
WAIC_stan <- T_stan+V_stan
U <- 0.5*log (2*pi*exp(1))+1/n/2
pre_AIC=AIC-3/n/2
## Graph 1
plot(AIC,WAIC_stan,col="red",pch=2,xlim=c(1.25,1.60),
     ylim=c(1.25,1.60),xlab="AIC",
     ylab="WAIC", main="AIC and WAIC")
abline(a=0,b=1,col="blue")
## Graph 2
plot(pre_AIC,T_stan,col="red",pch=3,xlim=c(1.25,1.60),
     ylim=c(1.25,1.60),xlab="AIC",
     ylab="WAIC", main="main terms of AIC and WAIC")
abline(a=0,b=1,col="blue")
## Graph 3
boxplot(G_stan,T_stan,names=c("G","T"),main="Boxplots of $G_n$, $T_n$ (n
     =200)")
abline(a=U,b=0,col="red")
```

It is possible to use the `waic` function of the CRAN `loo` package instead of constructing and calculating the `WAIC` function as above. However, in that case, the variable name `log_lik` must be used for the variable representing the log-likelihood. Also, the `waic` function displays the value of WAIC multiplied by $2n$ (twice the sample size).

Example 55 We applied the `WAIC` function and the `waic` function to the Boston data to find the value of WAIC. Also, since it is a multiple regression, we provide $x_{i,j}, i = 1, \ldots, n, j = 1, \ldots, p$ in the form of an $n \times (p + 1)$ design matrix X (the first column of X corresponds to the intercept and has all 1 values). The following processing was performed using `model11.stan` defined in Chap. 3. ∎

First, make the package and Boston data available.

```
library(rstan); library(MASS); library(loo); library(bayesplot)
data(Boston)
index <- c(1, 3, 5, 6, 7, 8, 10, 11, 12, 13, 14)
```

To obtain the design matrix X, execute the following.

```
1  lm <- formula(medv ~ . -medv, data=Boston)
2  df <- Boston[, index]
3  X=model.matrix(lm, df)
```

After that, it is no different from the usual Stan execution.

```
1  N <- nrow(df)
2  K <- length(index)
3  Y <- df$medv;
4  data_list <- list(N = N, M = K, y = Y, x = X)
5  fit <- stan(file = "model11.stan", data = data_list, seed = 1)
```

In the case of `waic` in the `loo` package, it can be calculated in the following two steps from `fit`.

```
1  m1 <- extract_log_lik(fit)
2  waic(m1)
```

As a result of the execution, the following output is obtained. is

```
Computed from 4000 by 506 log-likelihood matrix

            Estimate     SE
elpd_waic   -1536.4    33.8
p_waic         19.5     4.0
waic         3072.9    67.5
```

For the case of `WAIC`, it is as follows. However, we multiply by $2n$ to obtain the same value as `waic`.

```
1  m2 <- rstan::extract(fit)
2  2*N*WAIC(m2$log_lik)
```

As a result of the execution, the following output is obtained.

```
[1] 3072.8457
```

When comparing the values of WAIC for each statistical model, there is no need to multiply by $2n$.

6.4   Free Energy, BIC, and WBIC

Taking the logarithm of the marginal likelihood,

$$F_n := -\log Z_n$$

is called the free energy. When the true distribution is realizable and regular with respect to the statistical model, and when an appropriate prior probability is chosen, the free energy satisfies the property

$$\frac{1}{n} F_n \to \mathbb{E}_X[-\log p(X|\theta_*)].$$

The later described BIC (Bayesian Information Criterion) is obtained as an approximation of this free energy.

The value of the free energy is determined only by the samples. However, except for special cases such as when the statistical model belongs to the exponential family (Sect. 2.4) and the prior distribution is given by a conjugate prior distribution, it is difficult to calculate the exact value of the marginal likelihood, and therefore, the exact value of the free energy.

Example 56 For Example 4,

$$F_n = \frac{1}{2}\log(n+1) + \frac{1}{2}\sum_{i=1}^{n} x_i^2 - \frac{1}{2(n+1)}\left(\sum_{i=1}^{n} x_i\right)^2 + \frac{n}{2}\log 2\pi$$

is obtained. Then, we generate $x_1, \ldots, x_n$ ($n = 100$) according to the true distribution q (assuming the standard normal distribution) and measure the frequency of the free energy at that time (Fig. 6.5 left). We also tried a similar experiment for the binomial distribution (Fig. 6.5 right). The execution was done with the following code.

```
1   F.1 <-  function(x) log(n+1)/2+sum(x**2)/2-1/2/(n+1)*(sum(x))**2+n/2*log
2         (2*pi)
3   m <- 500
4   n <- 100
5   # Calculate free energy
6   T <- NULL
7   for(j in 1:m){
8     x <- rnorm(n)
9     T <- c(T,F.1(x))
10  }
11  # Plot the graph
12  plot(density(T), main="Free energy (Standard normal distribution)", col="
          red",
13        xlab="$F_n$", ylab="Probability density")
```

∎

In the following, we will consider cases where, like the above examples, the solution cannot be obtained analytically.

First, when the true distribution is regular with respect to the statistical model, from (5.8), the free energy and its average can be described as follows.

$$F_n = -\log Z_n = \sum_{i=1}^{n} -\log p(x_i|\theta_*) + \frac{d}{2}\log\frac{n}{2\pi}$$

$$+\frac{1}{2}\log \det J - \frac{n}{2}\Delta_n^\top J \Delta_n - \log \varphi(\theta_*) + o_P(1). \qquad (6.22)$$

If $x_1, \ldots, x_n \in \mathcal{X}$ are independent, we have

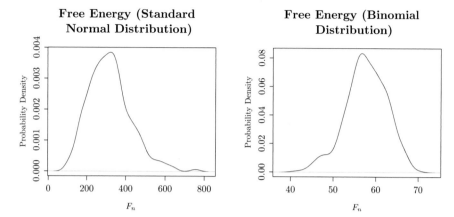

Fig. 6.5 In Example 4, the true distribution q was assumed to be the standard normal distribution. The free energy was generated 500 times, and its frequency was calculated (left). Next, in Example 5, the true distribution q was assumed to be a binomial distribution with probabilities 0.75 and 0.25 for 1 and 0, respectively. The free energy was generated 500 times, and its frequency was calculated (right)

$$\mathbb{E}_{X_1 \ldots X_n}[F_n]$$

$$= n \mathbb{E}_X[-\log p(X|\theta_*)] + \frac{d}{2} \log \frac{n}{2\pi} + \frac{1}{2} \log \det J - \frac{1}{2} \text{tr}(IJ^{-1}) - \log \varphi(\theta_*) + o(1)$$

where, in deriving the latter, (6.7) is used. In the realizable case, the average can be written as

$$\mathbb{E}_{X_1 \ldots X_n}[F_n] = n \mathbb{E}_X[-\log p(X|\theta_*)] + \frac{d}{2} \log \frac{n}{2\pi e} + \frac{1}{2} \log \det J - \log \varphi(\theta_*) + o(1).$$

$$(6.23)$$

From (5.10), (6.22) can also be written as

$$F_n = \sum_{i=1}^{n} -\log p(x_i|\hat{\theta}_n) + \frac{d}{2} \log \frac{n}{2\pi} + \frac{1}{2} \log \det J$$

$$- \frac{n}{2}(\hat{\theta}_n - \theta_* - \Delta_n)^\top J(\hat{\theta}_n - \theta_* - \Delta_n) - \log \varphi(\theta_*) + o_P(1)$$

$$= \sum_{i=1}^{n} -\log p(x_i|\hat{\theta}_n) + \frac{d}{2} \log \frac{n}{2\pi} + \frac{1}{2} \log \det J - \log \varphi(\theta_*) + o_P(1).$$

$$(6.24)$$

where we use the fact that the maximum likelihood estimate $\hat{\theta}_n = \theta_* + \Delta_n + o_P(1/\sqrt{n})$ (Proposition 20). The value obtained by subtracting the $O_P(1)$ term of the free energy (6.24) is called BIC (Bayesian Information Criteria).

$$BIC := \sum_{i=1}^{n} - \log p(x_i | \hat{\theta}_n) + \frac{d}{2} \log n.$$

There are several ways to approximate the free energy without assuming regularity. In statistical physics, the challenge has been to efficiently calculate the free energy with a small computational cost. As seen with WAIC, is there a method that can correctly calculate the free energy for general cases, even if they are not regular, while having performance similar to BIC when regular? In the following, we will consider the information criterion

$$WBIC := \mathcal{E}_\beta [\sum_{i=1}^{n} - \log p(x_i | \theta)] , \tag{6.25}$$

where we have set

$$\mathcal{E}_\beta[f(\theta)] := \int_\Theta f(\theta) p_\beta(\theta | x_1, \ldots, x_n) d\theta,$$

with

$$p_\beta(\theta | x_1, \ldots, x_n) = \frac{\varphi(\theta) \prod_{i=1}^{n} p(x_i | \theta)^\beta}{Z_n(\beta)}$$

and

$$Z_n(\beta) := \int_\Theta \varphi(\theta) \prod_{i=1}^{n} p(x_i | \theta)^\beta d\theta.$$

WBIC has various desirable properties.

Proposition 26 *In particular, when $\beta = 1/\log n$ is set, when the true distribution is regular with respect to the statistical model, we have*

$$WBIC = BIC + o_P(1). \tag{6.26}$$

Proof Refer to the appendix at the end of the chapter.

Even if it is not regular, when $\beta = 1/\log n$, in general,

$$F_n = WBIC + O_P(\sqrt{\log n})$$

holds (proof in Chap. 7). In any case, it provides an efficient method for calculating the free energy.

To write WBIC in Stan, it would look like the following.

```
1  wbic <- function(log_likelihood) - mean(rowSums(log_likelihood))
```

Example 57 For the p-variable multiple regression (with intercept), we examined the relationship in (6.26).

$$f(x|\mu) = \frac{1}{\sqrt{2\pi\sigma^2}} \exp\{-\frac{(x-\mu)^2}{2\sigma^2}\}.$$

By setting the partial derivatives of

$$L := -\sum_{i=1}^{n} \log f(x_i|\mu, \sigma^2) = \frac{n}{2} \log(2\pi\sigma^2) + \frac{1}{2\sigma^2} \sum_{i=1}^{n} (x_i - \mu)^2$$

with respect to μ and σ^2 to 0, the maximum likelihood estimates $\hat{\mu} = \frac{1}{n} \sum_{i=1}^{n} x_i, \hat{\sigma}^2 = \frac{1}{n} \sum_{i=1}^{n} (x_i - \hat{\mu})^2$ are obtained. Therefore, substituting the maximum likelihood estimates $\hat{\mu}, \hat{\sigma}^2$ into L, we have

$$L = \frac{n}{2} \log(2\pi\hat{\sigma}^2) + \frac{1}{2\hat{\sigma}^2} \sum_{i=1}^{n} (x_i - \hat{\mu})^2 = \frac{n}{2} \log(2\pi\hat{\sigma}^2 e)$$

and

$$BIC = \frac{n}{2} \log(2\pi\hat{\sigma}^2 e) + \frac{p+1}{2} \log n.$$

The execution was done using the R code below. Also, the Stan file for WBIC (model15.stan) was configured as shown below. We conducted experiments $m = 20$ times each for $p = 3, 7$ and $n = 100, 300, 500$, and obtained the results in Fig. 6.6. ∎

```
1   library(rstan)
2   bic <- function(x,y){
3       beta2 <- as.vector(solve(t(x)%*%x)%*%t(x)%*%y)
4       sigma2 <- sum((y-x%*%beta2)^2)/n
5       return(0.5*n*log(2*pi*exp(1)*sigma2)+0.5*(p+1)*log(n))
6   }
7   model15 <- stan_model("model15.stan")
8   m <- 20
9   n <- 200
10  p <- 3 ones
11  <- rep(1,n)
12  BIC <- NULL
13  WBIC <- NULL
14  for(j in 1:m){
15      beta <- rnorm(p+1)
16      x <- matrix(rnorm(n*p),n,p)
17      x <- cbind(ones,x)
```

```
18    y <- as.vector(x%*%beta+rnorm(n))
19    BIC <- c(BIC,bic(x,y))
20    data_list <- list(N = n, M = p+1, y = y, x = x, beta=1/log (n))
21    fit <- sampling(model15, data = data_list, seed = 1,iter=3000)
22    mm <- rstan::extract(fit)
23    WBIC <- c(WBIC, wbic(mm$log_lik))
24  }
25  plot(BIC,WBIC, xlab="BIC",ylab="WBIC",xlim=c(),)
```

In the case of multiple regression, the value of WBIC can be obtained by preparing the following Stan code. First, input the inverse temperature $\beta > 0$. Then, pay attention to the point where the log posterior likelihood is determined based on that β. If $\beta = 1$, it becomes

```
y ~ normal(X * b, sigma);
```

or

```
for(n in 1:N)
    target += normal_lpdf(Y[n] | X[n] * b, sigma);
```

If $\beta \neq 1$, it means that the definition of the posterior probability itself is different. Also, it is difficult to model with the former expression, including β. In that sense, the sampling itself is different. Of course, we must multiply β in the `model` block's `target`, not the `generated quantities` block's `log_lik`.

model15.stan

```
1  data {
2    int N;              // Sample size
3    int M;              // Number of explanatory variables + 1 (intercept
```

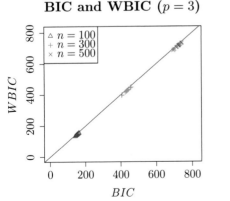
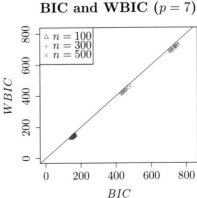

Fig. 6.6 We compared the values of BIC and WBIC for multiple regression ($p = 3, 7$ variables, including the intercept) when $n = 100, 300, 500$. Since it is a regular statistical model, it can be seen that the values of both are almost the same in both cases of $p = 3, 7$

```
4      )
5      vector[N] y;              // Dependent variable
6      matrix[N, M] x;           // Design matrix
7      real beta;                // Inverse temperature
8    }
9    parameters {
10     vector[M] b;              // M−1 slopes and intercept
11     real<lower=0> sigma;  // Standard deviation
12   }
13   model {
14     for(n in 1:N)
15       target += beta * normal_lpdf(y[n] | x[n] * b, sigma);
16   }
17   generated quantities{
18     vector[N] log_lik;
19     for (n in 1:N)
20       log_lik[n]= normal_lpdf(y[n] | x[n]*b, sigma);
21   }
```

In this book, for a concise understanding, we take a policy of not using the inverse temperature $\beta > 0$ until the second half (it is easier to understand without being aware of $\beta > 0$). However, if we introduce the inverse temperature $\beta > 0$, we will have the same posterior probability structure even without considering specific applications such as WAIC and WBIC. In fact, model15.stan will be applied to WAIC and cross-validation, assuming a general inverse temperature $\beta > 0$ in Chap. 8.

Appendix: Proof of Proposition

Proof of Proposition 20

To improve the visibility, we shall discuss for a general inverse temperature $\beta > 0$. $\beta \to \infty$ corresponds to the maximum likelihood estimation case.

Since $\eta_n(\theta)$ converges in probability to a certain random variable for each $\theta \in \Theta$ (Proposition 10) and any $\beta > 0$,

$$
\begin{aligned}
\mathcal{L}(\theta) &:= -\frac{1}{n} \sum_{i=1}^{n} \log p(x_i|\theta) - \frac{1}{n\beta} \log \varphi(\theta) \\
&= -\frac{1}{n} \sum_{i=1}^{n} \log p(x_i|\theta_*) + \mathbb{E}_X[\log \frac{p(X|\theta_*)}{p(X|\theta)}] - \frac{1}{\sqrt{n}} \eta_n(\theta) - \frac{1}{n\beta} \log \varphi(\theta) \\
&= -\frac{1}{n} \sum_{i=1}^{n} \log p(x_i|\theta_*) + \mathbb{E}_X[\log \frac{p(X|\theta_*)}{p(X|\theta)}] + O_P(\frac{1}{\sqrt{n}}).
\end{aligned}
\tag{6.27}
$$

Then, for $\epsilon_n = n^{-1/4}$, $\epsilon_n \to 0$ and $\sqrt{n}\epsilon_n \to \infty$ $(n \to \infty)$, for $\theta \in \Theta$ such that $\mathbb{E}_X[\log \frac{p(X|\theta_*)}{p(X|\theta)}] \geq \epsilon_n$,

$$\mathcal{L}(\theta) \geq -\frac{1}{n}\sum_{i=1}^{n} \log p(x_i|\theta_*) + \epsilon_n + O_P(\frac{1}{\sqrt{n}}).$$

The right side becomes larger than

$$\mathcal{L}(\theta_*) = -\frac{1}{n}\sum_{i=1}^{n} \log p(x_i|\theta_*) + O_P(\frac{1}{\sqrt{n}})$$

when $n \to \infty$. Therefore, for θ $(\hat{\theta})$ that minimizes $\mathcal{L}(\theta)$, it is necessary that the probability of $\mathbb{E}_X[\log \dfrac{p(X|\theta_*)}{p(X|\hat{\theta})}] < \epsilon_n$ converges to 1. Moreover, since it is regular and θ_* is unique, $\hat{\theta}$ converges in probability to θ_*. Also, the same holds when β is arbitrarily large, so the first half of the claim is demonstrated.

Regarding the second half, since $\nabla\mathcal{L}(\hat{\theta}) = 0$, by the mean value theorem, there exists a θ^1 between θ_* and $\hat{\theta}$ such that

$$0 = \nabla\mathcal{L}(\theta_*) + \nabla^2\mathcal{L}(\theta^1)(\hat{\theta} - \theta_*)$$

and we have
$$\hat{\theta} = \theta_* - \nabla^2\mathcal{L}(\theta^1)^{-1}\nabla\mathcal{L}(\theta_*). \tag{6.28}$$

Next, differentiating (6.27) twice with respect to θ and substituting $\theta = \theta^1$, from the first half of the claim $(\hat{\theta} \xrightarrow{p} \theta_*)$,

$$\nabla^2\mathcal{L}(\theta^1) = \nabla^2\mathbb{E}_X[-\log p(X|\theta)]\big|_{\theta=\theta^1} + o_P(1) = J + o_P(1) \tag{6.29}$$

holds. Finally, differentiating the definition of $\mathcal{L}(\theta)$ once with respect to θ and substituting $\theta = \theta_*$, from (5.1) and (5.2),

$$-\nabla\mathcal{L}(\theta_*) = \frac{1}{\sqrt{n}}\nabla\eta_n(\theta_*) + o_P(\frac{1}{\sqrt{n}}) = J\Delta_n + o_P(\frac{1}{\sqrt{n}}) \tag{6.30}$$

holds (actually, $o_P(\frac{1}{\sqrt{n}})$ becomes $O_P(\frac{1}{n})$). Therefore, from (6.28), (6.29), and (6.30), we obtain the second half of the claim

$$\hat{\theta} = \theta_* + (J + o_P(1))^{-1}\{J\Delta_n + o_P(\frac{1}{\sqrt{n}})\} = \theta_* + \Delta_n + o_P(\frac{1}{\sqrt{n}}).$$

∎

Proof of Proposition 26

Proof From Proposition 13,[8] we can write

$$\mathcal{E}_\beta[\sum_{i=1}^n \log \frac{p(x_i|\theta_*)}{p(x_i|\theta)}] = \frac{\int_\Theta \{\sum_{i=1}^n \log \frac{p(x_i|\theta_*)}{p(x_i|\theta)}\}\{\prod_{i=1}^n p(x_i|\theta)\}^\beta \varphi(\theta)d\theta}{\int_\Theta \{\prod_{i=1}^n p(x_i|\theta)\}^\beta \varphi(\theta)d\theta}$$

$$= \frac{U_n^{(3)} + o_P(\exp(-\sqrt{n}))}{U_n^{(1)} + o_P(\exp(-\sqrt{n}))}.$$

where we define

$$U_n^{(1)} := \int_{B(\epsilon_n,\theta_*)} \{\prod_{i=1}^n \frac{p(x_i|\theta)}{p(x_i|\theta_*)}\}^\beta \varphi(\theta)d\theta$$

and

$$U_n^{(3)} := \int_{B(\epsilon_n,\theta_*)} \{\sum_{i=1}^n \log \frac{p(x_i|\theta_*)}{p(x_i|\theta)}\}\{\prod_{i=1}^n \frac{p(x_i|\theta)}{p(x_i|\theta_*)}\}^\beta \varphi(\theta)d\theta.$$

Next, by the mean value theorem, there exists a $\theta^1 \in \Theta$ between θ and $\hat{\theta}$ such that

$$\sum_{i=1}^n -\log p(x_i|\theta) = \sum_{i=1}^n -\log p(x_i|\hat{\theta}) + \frac{1}{2}(\theta - \hat{\theta})^\top J_n(\theta^1)(\theta - \hat{\theta}) \qquad (6.31)$$

where we define the (j, k)-th component of $J_n(\theta)$ is defined as $-\sum_{i=1}^n \frac{1}{n}\frac{\partial^2 \log p(x_i|\theta)}{\partial\theta_j\partial\theta_k}$.

Furthermore, from the probability convergence $\hat{\theta} \to \theta_*$ and $\epsilon_n = n^{-1/4}$, the probability convergence $\theta^1 \to \theta_*$ holds. Therefore, taking the Frobenius norm[9] as $|\cdot|$, as $n \to \infty$, the right-hand side of

$$\|J_n(\theta^1) - J(\theta_*)\| \le \|J_n(\theta^1) - J_n(\theta_*)\| + \|J_n(\theta_*) - J(\theta_*)\|$$

$$\le |\theta_1 - \theta_*|^\top \sup_{\theta \in B(\epsilon_n,\theta_*)} \|\frac{\partial J_n(\theta)}{\partial\theta}\| + \|J_n(\theta_*) - J(\theta_*)\|$$

converges to 0. Thus,

$$J_n(\theta^1) = J(\theta_*) + o_P(1) \qquad (6.32)$$

---

[8] It is proven for $\beta = 1$, but it holds for a general β.

[9] The square root of the sum of squares of all components. It is known to satisfy the norm conditions.

holds. Since it is assumed to be regular, $J(\theta_*)$ is positive definite. Hence, from (6.31) and (6.32), similarly to the proof of Proposition 14, we can write

$$U_n^{(1)} = \{\prod_{i=1}^{n} \frac{p(x_i|\hat{\theta})}{p(x_i|\theta_*)}\}^{\beta} \int_{B(\epsilon_n,\theta_*)} \exp\{-\frac{n\beta}{2}(\theta-\hat{\theta})^{\top}(J(\theta_*)+o_P(1))(\theta-\hat{\theta})\}\varphi(\theta)d\theta$$

$$= \{\prod_{i=1}^{n} \frac{p(x_i|\hat{\theta})}{p(x_i|\theta_*)}\}^{\beta}(n\beta)^{-d/2} \int_{\mathbb{R}^d} \exp\{-\frac{1}{2}u^{\top}(J(\theta_*)+o_P(1))u\}\varphi(\hat{\theta}+\frac{u}{\sqrt{n\beta}})du$$

$$\sim \{\prod_{i=1}^{n} \frac{p(x_i|\hat{\theta})}{p(x_i|\theta_*)}\}^{\beta} \left(\frac{2\pi}{n\beta}\right)^{d/2} \frac{\varphi(\hat{\theta})+o_P(1)}{\sqrt{\det(J(\theta_*)+o_P(1))}}.$$

where we define $u = \sqrt{n\beta}(\theta-\hat{\theta})$. Furthermore,

$$U_n^{(3)} = \{\prod_{i=1}^{n} \frac{p(x_i|\hat{\theta})}{p(x_i|\theta_*)}\}^{\beta} \int_{B(\epsilon_n,\theta_*)} \{\sum_{i=1}^{n} -\log p(x_i|\hat{\theta}) + \frac{n\beta}{2}(\theta-\hat{\theta})^{\top}(J(\theta_*)+o_P(1))(\theta-\hat{\theta})\}$$

$$\cdot \exp\{-\frac{n\beta}{2}(\theta-\hat{\theta})^{\top}(J(\theta_*)+o_P(1))(\theta-\hat{\theta})\}\varphi(\theta)d\theta$$

$$\sim \{\prod_{i=1}^{n} \frac{p(x_i|\hat{\theta})}{p(x_i|\theta_*)}\}^{\beta} \left(\frac{2\pi}{n\beta}\right)^{d/2} \frac{\varphi(\hat{\theta})+o_P(1)}{\sqrt{\det(J(\theta_*)+o_P(1))}} \{\sum_{i=1}^{n} -\log p(x_i|\hat{\theta}) + \frac{d}{2\beta} + o_P(1)\}$$

can be written. Since $\beta = 1/\log n$, these imply Proposition 26.

■

Exercises 54–66

In the following, let $m, n \geq 1$.

54. When $I_n = J_n$, show that the second term of (6.3) becomes d/n.
55. Show that the minimum value of the negative log-likelihood in Example 50 is given by (6.4).
56. Explain the relationship between the input X, y, T and the output S.min, set.q of the function RSS.min. Also, identify which variables are used as global variables in the function IC.
57. In the proof of Proposition 20, for any $\beta > 0$, let

$$\mathcal{L}(\theta) := -\frac{1}{n}\sum_{i=1}^{n}\log p(x_i|\theta) - \frac{1}{n\beta}\log\varphi(\theta)$$

be minimized by θ (denoted as $\hat{\theta}$). Show that the probability of $\mathbb{E}[\log \dfrac{p(X|\theta_*)}{p(X|\hat{\theta})}] <$ ϵ_n converges to 1. Also, show that there exists a θ^1 between θ_* and $\hat{\theta}$ such that

$$\hat{\theta} = \theta_* - \nabla^2 \mathcal{L}(\theta^1)^{-1} \nabla \mathcal{L}(\theta_*)$$

Moreover, show $\nabla^2 \mathcal{L}(\theta^1) = J + o_P(1)$ and $-\nabla \mathcal{L}(\theta_*) = J \Delta_n + o_P(\frac{1}{\sqrt{n}})$ hold.

58. When $X_1, \ldots, X_m \sim N(0, 1)$ are independent, show the following. It is allowed to use that their square mean is 1 and the fourth-power mean is 3 (proven in Chap. 1).。

   (a) The mean of $X_1^2, \ldots, X_m^2$ is 1, and the variance is 2.
   (b) The mean of $X_1^2 + \cdots + X_m^2$ is m, and the variance is $2m$.

59. Prove each of the following $\Longrightarrow$.

   (a) Propositions 20 and 21 $\Longrightarrow$ Proposition 22
   (b) Propositions 9 and 22 $\Longrightarrow$ Proposition 23
   (c) Propositions 19 and 21 $\Longrightarrow$ Proposition 24

60. Why can the WAIC value defined in (6.20) be calculated from the function T_n for empirical loss in Chap. 4 and the following function?

```
1  V_n <- function(log_likelihood) mean(colMeans(log_likelihood^2) -
       colMeans(log_likelihood)^2)
2  WAIC <- function(log_likelihood) T_n (log_likelihood) + V_n (log_
       likelihood)
```

61. For the true distribution and statistical model of the Gaussian mixture distribution in Example 48, calculate the values of AIC and WAIC.

62. For the continuous Boston data in Example 55, calculate the WAIC values using both the function WAIC and the function waic from the loo package, and confirm that the values match.

63. In the case where the true distribution is regular with respect to the statistical model, show that the free energy can be expressed as (6.22). Also, using Jeffreys' prior probability

$$\varphi(\theta) = \frac{\sqrt{\det J(\theta)}}{\int_\Theta \sqrt{\det J(\theta')} d\theta'}$$

show that the mean (6.23) can be written as

$$n \mathbb{E}_X[-\log p(X|\theta_*)] + \frac{d}{2} \log \frac{n}{2\pi e} + \log \int_\Theta \sqrt{\det J(\theta)} d\theta + o(1)$$

when realizable.

64. Following the graph of the free energy of the standard normal distribution in Example 56, for Example 5, generate a length $n = 100$ binary vector with each value independently having a probability $p = 0.25$ of occurring as 1, and perform the operation of calculating its free energy $m = 500$ times to generate a graph.

65. Explain how WBIC calculations can be implemented using the following code.

```
1  wbic <- function(log_likelihood) - mean(rowSums(log_likelihood))
```

66. Replace the data in Example 57 with continuous Boston data and compare the values of BIC and WBIC.

Chapter 7
Algebraic Geometry

In this chapter, we will study algebraic geometry and its surroundings. First, we will learn about algebraic sets and manifolds. A manifold is like a globe (an open set family) that has many local maps (open sets). Each open set must be a one-to-one correspondence with the open set of the same dimensional Euclidean space, which allows us to define local variables and local coordinates. The resolution of singular points, referred to as blow-ups, denotes the process of updating local coordinates containing singular points to other local coordinates. The Watanabe-Bayes theory aims to obtain a standard form called normal crossing for each local coordinate. In the regular case, the dimension d of the parameter is twice the real logarithmic threshold λ in the general case. This value of λ can be obtained by resolving singular points. In fact, the resolution of singular points is not directly related to the Watanabe-Bayes theory. This point is often misunderstood in Watanabe-Bayes theory. Based on the Hironaka theorem, whether there are singular points or not, in each local coordinate, we transform the average log-likelihood to normal crossing.

Readers who are learning algebraic geometry for the first time may not understand what is written here at all. In such a case, as mentioned in the "Introduction", I recommend slowly reading while writing the formulas in each section. If you still do not understand, I recommend repeating the same thing tomorrow and the day after. Eventually, you should feel more comfortable.

7.1 Algebraic Sets and Analytical Sets

Hereafter, we denote the set of real-number-coefficient polynomials with variables $x = (x_1, \ldots, x_d)$ as $\mathbb{R}[x_1, \ldots, x_d]$ or $\mathbb{R}[x]$. At this time, using the subset J of $\mathbb{R}[x]$ (assuming it is not an empty set), we define the set I that can be written as

J. Suzuki, *WAIC and WBIC with R Stan*,
https://doi.org/10.1007/978-981-99-3838-4_7

$$I = \left\{ \sum_i f_i(x) g_i(x) \mid f_i(x) \in J, \, g_i(x) \in \mathbb{R}[x] \right\}$$

as the *ideal* of $\mathbb{R}[x]$. We also say that set J generates the ideal I.

Example 58 For $J = \{x + y^2, x - y^2, y^3\} \subseteq \mathbb{R}[x, y]$, $I = \{x f(x, y) + y^2 g(x, y) \mid f, g \in \mathbb{R}[x, y]\}$ is the ideal generated by J. In other words, the same ideal is generated by $J = \{x, y^2\}$. While $I_1 = \{x f(x, y) \mid f \in \mathbb{R}[x, y]\}$ and $I_2 = \{y^2 g(x, y) \mid g \in \mathbb{R}[x, y]\}$ are ideals, we have

$$I \ni x - y^2 \notin I_1 \cup I_2.$$

∎

The set of common zeros of the elements of the ideal I is given by

$$V(I) = \left\{ x \in \mathbb{R}^d \mid f(x) = 0, \, f \in I \right\} \subseteq \mathbb{R}^d$$

and it is called the *algebraic set* determined by the ideal I in $\mathbb{R}[x]$. If the algebraic set V can be expressed as the union of two distinct non-empty algebraic sets V_1 and V_2, it is said to be *reducible*. Otherwise, it's called *irreducible*. Hereafter, we assume that $V(I)$ is irreducible, and for simplicity, we will refer to it simply as V. In contrast with the forthcoming projective space $\mathbb{P}^d$, we sometimes denote the d-dimensional Euclidean space $\mathbb{R}^d$ as the affine space $\mathbb{A}^d$ according to tradition. When given the algebraic set V, note that the subset of $\mathbb{R}[x]$

$$I(V) = \{f \in \mathbb{R}[x] \mid f(x) = 0, x \in V\}$$

forms an ideal in $\mathbb{R}[x]$.

Example 59 The algebraic set of the ideal I generated by $J = \{x^2 + y^2\}$ is

$$V = V(I) = \left\{ (x, y) \in \mathbb{R}^2 \mid x^2 + y^2 = 0 \right\} = \{(0, 0)\}$$

Consequently, $I(V)$ is

$$I(V) = \{f \in \mathbb{R}[x, y] \mid f(0, 0) = 0\} = \left\{ x, y, x^2, y^2, xy, \ldots \right\}$$

and it becomes the ideal generated by $J = \{x, y\}$.

Similarly, we can construct an *analytic set*

$$\{x \in U \mid f(x) = 0\}$$

$$a = -3, b = 3$$

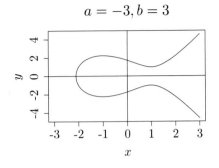

$$a = 1, b = 0$$

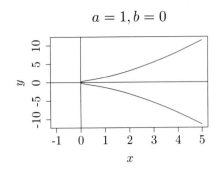

$$a = -1, b = 0$$

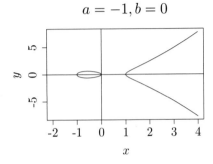

Fig. 7.1 The case where the elliptic curve $y^2 = x^3 + ax + b$ does not have a singularity (refer to Sect. 6.3). The above three types are considered typical

using an analytic function f with an open set $U \subseteq \mathbb{R}^d$ as its domain. Using multiple analytic functions $f_1, \ldots, f_d : U \to \mathbb{R}$ that share the domain U, we can also define an analytic set

$$\{x \in U \mid f_1(x) = 0, \ldots, f_d(x) = 0\}.$$

Also, if $f : U \to V$ is defined as $f = (f_1, \ldots, f_d)$ by analytic functions $f_1, \ldots, f_d : U \to \mathbb{R}$ with U, V being open sets in $\mathbb{R}^d$, we call f an *analytic map*.

In this chapter, we mainly focus on algebraic sets defined by a single irreducible (i.e., cannot be factored further) polynomial $0 \neq f \in \mathbb{R}[x]$

$$V(f) = \{x \in \mathbb{R}^d \mid f(x) = 0\} \subseteq \mathbb{R}^d$$

or analytic sets constituted by a single analytic function $f : U \to \mathbb{R}$

$$V(f) = \{x \in U \mid f(x) = 0\} \subseteq U.$$

Example 60 *(Elliptic curve)* For $a, b \in \mathbb{R}$, a curve on a plane determined by the polynomial of 2 variables $f(x, y) = y^2 - x^3 - ax - b = 0$ is called an elliptic curve. Try to draw the outline of the algebraic set

$$V(f) = \{(x, y) \in \mathbb{R}^2 \mid f(x, y) = 0\}$$

for $(a, b) = (-3, 3), (1, 0), (-1, 0)$ using the R language (Fig. 7.1). The following code is used: ∎

```
1   a <- -3
2   b <- 3
3   x.min <- -3
4   x.max <- 3        # Non-singular case (1)
5   # a <- 1; b <- 0; x.min <- -1; x.max <- 5       # Non-singular case (2)
6   # a <- -1; b <- 0; x.min <- -2; x.max <- 4      # Non-singular case (3)
7   # a <- 0; b <- 0; x.min <- -1; x.max <- 5       # Case including singularity
8       (1)
9   # a <- -3; b <- 2; x.min <- -3; x.max <- 3      # Case including singularity
10      (2)
11  f <- function(x) sqrt(max(x^3+a*x+b,0))
12  x.seq <- seq(x.min,x.max,0.001)
13  y.seq <- NULL
14  for(x in x.seq) y.seq <- c(y.seq,f(x))
15  y.max <- max(y.seq)
16  plot(0,xlab="x", ylab="y",xlim=c(x.min,x.max), ylim=c(-y.max,y.max),type=
17      "n",
18      main=paste("a=",a,", b=",b))
19  lines(x.seq,y.seq)
20  lines(x.seq,-y.seq)
21  abline(h=0)
22  abline(v=0)
```

7.2 Manifold

In this section, we define *topological spaces* and (analytic) *manifolds*. Let M be a set. When a set $\mathcal{U}$ consisting of subsets of M is defined to satisfy the following three conditions, M is called a topological space, $\mathcal{U}$ is called a *family of open sets*, and the elements of $\mathcal{U}$ are called open sets.[1]

1. $\mathcal{U}$ includes the entire set M and the empty set $\{\}$ as elements.
2. The union of any number of elements of $\mathcal{U}$ (open sets of M) is an element of $\mathcal{U}$.
3. The intersection of any finite number of elements of $\mathcal{U}$ (open sets of M) is an element of $\mathcal{U}$.

Furthermore, for any $x \neq y \in M$, when there exist $U, V \in \mathcal{U}$ such that $x \in U, y \in V$, and $U \cap V = \{\}$, that topological space M is called a *Hausdorff space*.

Example 61 *(Distance space)* If a distance $d(x, y), x, y \in M$ is defined for the set M, we can define an open set $B(\epsilon, x) := \{y \in M \mid dist(x, y) < \epsilon\}$ using it, so the family of open sets can be defined as

---

[1] Including distance metrics such as Euclidean distance to define open sets (metric spaces).

Fig. 7.2 The images of
$\phi_1(U_1 \cap U_2)$ and
$\phi_2(U_1 \cap U_2)$ by ϕ_1 and ϕ_2
are connected by $\phi_2 \circ \phi_1^{-1}$
and $\phi_1 \circ \phi_2^{-1}$ (coordinate
transformation), which are
analytic mappings

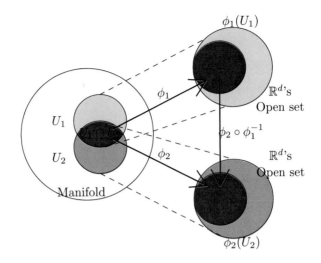

$$\mathcal{U} = \{B(\epsilon, x) \mid \epsilon > 0, x \in M\}.$$

For $x, y \in M$, $x \neq y$, by taking $\epsilon > 0$ sufficiently small, we can make $B(\epsilon, x) \cap B(\epsilon, y) = \{\}$, so if the topological space M is a distance space, it is Hausdorff. ∎

In the following, we define a manifold.[2] Let M be a Hausdorff topological space. When a bijection (mapping one-to-one and onto) from an open set to another open set is continuous in both directions, this mapping is said to be *homeomorphic*.

1. For each open set U of the family of open sets of M, there exists a ϕ such that $U \to \phi(U) \subseteq \mathbb{R}^d$ is homeomorphic.
2. For such pairs (U, ϕ) and $(\tilde{U}, \tilde{\phi})$, when $U \cap \tilde{U}$ is not empty, the *coordinate transformation*

$$\phi \circ \tilde{\phi}^{-1}(U \cap \tilde{U}) : \phi(U \cap \tilde{U}) \to \tilde{\phi}(U \cap \tilde{U})$$

is an analytic mapping (see Fig. 7.2).

In this case, M is said to be a d-dimensional analytic manifold. At this time, each element u of U can be treated as if it were an element $\phi(u)$ of the open set $\phi(U)$ of $\mathbb{R}^d$. This $\phi(u) \in \mathbb{R}^d$ is called a *local variable*, and the coordinates constructed by them are called *local coordinates*. By using local coordinates, it is possible to treat a point on U as if it were a point in $\mathbb{R}^d$. In addition, such a set consisting of (U, ϕ) is called a *local coordinate system* of M.

Example 62 $\mathbb{A}^d$ is a (trivial) d-dimensional manifold. With the identity mapping $id : \mathbb{R}^d \to \mathbb{R}^d$, $S = \{(\mathbb{R}^d, id)\}$ forms a coordinate neighborhood system. It may be

---

[2] A topological space being Hausdorff is a necessary condition for the existence of a partition of unity (Sect. 7.5)

divided into multiple local coordinates. When $d = 1$, define $U_i = (i - 1, i + 1)$ and ϕ_i for each $i \in \mathbb{Z}$ as

$$\phi_i : U_i \ni x \mapsto x - i \in (-1, 1).$$

Then, ϕ_i becomes a local coordinate system of U_i, and $S = \{(U_i, \phi_i) \mid i \in \mathbb{Z}\}$ gives a coordinate neighborhood system. In this case, for $x \in (0, 1)$,

$$\phi_{i+1} \circ \phi_i^{-1}(x) = x - 1$$

becomes the coordinate transformation. And for each i, $x - i \in (-1, 1)$ can be used as a local variable.   ∎

As a typical example of a manifold, we consider the projective space. For each $(x_0, x_1, \ldots, x_d), (x_0', x_1', \ldots, x_d') \in \mathbb{R}^{d+1} \setminus \{(0, \ldots, 0)\}$, when there exists a $t \in \mathbb{R}$ such that

$$(x_0, x_1, \ldots, x_d) = t(x_0', x_1', \ldots, x_d') ,$$

an equivalence relationship exists between them. When each class is denoted as $[x_0 : x_1 : \cdots : x_d]$, the set of elements is written as $\mathbb{P}^d$, which is called a d-*dimensional projective space*.

Example 63 $\mathbb{P}^d$ is a d-dimensional manifold. When the value of the ith coordinate is not zero, by dividing the values of other coordinates by its value, we get $U_i := \{[x_0 : x_1 : \cdots : x_{i-1} : 1 : x_{i+1} : \cdots : x_d] \in \mathbb{P}^d\}$. The coordinate transformation

$$\phi_i : U_i \ni [x_0 : x_1 : \cdots : x_{i-1} : 1 : x_{i+1} : \cdots : x_d] \mapsto (x_0, x_1, \ldots, x_{i-1}, x_{i+1}, \ldots, x_d) \in \mathbb{A}^d$$

from $\phi_i(U_i \cap U_j)$ to $\phi_j(U_i \cap U_j)$ becomes

$$
\phi_j \circ \phi_i^{-1}(x_0, x_1, \ldots, x_{i-1}, x_{i+1}, \ldots, x_d)
$$
$$
= \begin{cases} \left(\dfrac{x_0}{x_j}, \dfrac{x_1}{x_j}, \ldots, \dfrac{x_{i-1}}{x_j}, \dfrac{1}{x_j}, \dfrac{x_{i+1}}{x_j}, \ldots, \dfrac{x_{j-1}}{x_j}, \dfrac{x_{j+1}}{x_j}, \ldots, \dfrac{x_d}{x_j} \right), & i < j \\ \left(\dfrac{x_0}{x_j}, \dfrac{x_1}{x_j}, \ldots, \dfrac{x_{j-1}}{x_j}, \dfrac{x_{j+1}}{x_j}, \ldots, \dfrac{x_{i-1}}{x_j}, \dfrac{1}{x_j}, \dfrac{x_{i+1}}{x_j}, \ldots, \dfrac{x_d}{x_j} \right), & j < i \end{cases} , \quad (7.1)
$$

(Exercise 70(a)) and $S = \{(U_i, \phi_i) \mid i = 0, 1, \ldots, d\}$ forms a coordinate neighborhood system. And for each i,

$$(x_0, x_1, \ldots, x_{i-1}, x_{i+1}, \ldots, x_d) \in \mathbb{R}^d$$

can be used as local coordinates. Particularly, in the case of $d = 1$, it becomes

$$\phi_x : U_x \ni [1 : u_x] \mapsto u_x \in \mathbb{A}^1$$

and

$$\phi_y : U_y \ni [u_y : 1] \mapsto u_y \in \mathbb{A}^1.$$

And when $u_x, u_y \neq 0$, the coordinate transformation is given by (Exercise 70(b))

$$\phi_{xy} : \phi_x(U_x \cap U_y) \ni u_x \mapsto \frac{1}{u_x} = u_y \in \phi_y(U_x \cap U_y)$$

and

$$\phi_{yx} : \phi_y(U_x \cap U_y) \ni u_y \mapsto \frac{1}{u_y} = u_x \in \phi_x(U_x \cap U_y)$$

from $[1 : u_x] = [u_y : 1]$.   ∎

When looking at a certain country on a globe, one cannot see the country on the other side of the earth unless the globe is rotated. It may be interpreted that the globe is made by pasting together multiple maps.[3]

7.3 Singular Points and Their Resolution

Next, we define singular points on an algebraic set V. If the ideal $I(V)$ is generated by polynomials $f_1, \ldots, f_m$, and the rank of the matrix

$$\left(\frac{\partial f_i(x_1, \ldots, x_d)}{\partial x_j} \right)_{i=1,\ldots,m, j=1,\ldots,d}$$

is constant for every $x = (x_1, \ldots, x_d) \in V$, then V is said to be non-singular. If there exists an $x \in V$ where the rank is smaller, then x is called a singular point. Especially when $m = 1$ and

$$f_1(x_1, \ldots, x_d) = \frac{\partial f_1}{\partial x_1}(x_1, \ldots, x_d) = \cdots = \frac{\partial f_1}{\partial x_d}(x_1, \ldots, x_d) = 0, \qquad (7.2)$$

$(x_1, \ldots, x_d) \in V$ is called a singular point of V. If V has no singular points, it's said to be non-singular.

Example 64 From Example 59, $I(V)$ is generated by $J = \{x, y\}$, thus

$$\begin{bmatrix} \frac{\partial f_1(x,y)}{\partial x} & \frac{\partial f_1(x,y)}{\partial y} \\ \frac{\partial f_2(x,y)}{\partial x} & \frac{\partial f_2(x,y)}{\partial y} \end{bmatrix} = \begin{bmatrix} 1 & 0 \\ 0 & 1 \end{bmatrix}$$

holds for all $(x, y) \in V$. Therefore, V is non-singular.

---

[3] In this sense, manifolds are sometimes called "atlases", and individual (U_i, ϕ_i) are called "charts".

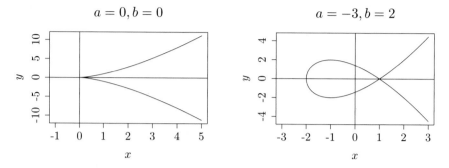

Fig. 7.3 When the elliptic curve $y^2 = x^3 + ax + b$ has a singular point. In the case of $(a, b) = (0, 0)$ (left), $(x, y) = (0, 0)$ becomes a singular point, and in the case of $(a, b) = (-3, 2)$ (right), $(x, y) = (1, 0)$ becomes a singular point. In each case, it can be seen that a tangent line cannot be drawn (as it is sharp), and multiple tangent lines can be drawn. Conversely, a non-singular algebraic curve can draw a single tangent line at any point, and can be said to be smooth

Example 65 *(Singular elliptic curve)* In Example 60, we look for the condition for having a singular point. As (7.3) specifically becomes

$$y^2 = x^3 + ax + b \ , \ 3x^2 + a = 0 \ , \ 2y = 0$$

$$\Longleftrightarrow x(-\frac{a}{3} + a) + b = 0 \ , \ 3x^2 + a = 0 \ , \ y = 0$$

$$\Longleftrightarrow \begin{cases} a = b = 0, & (x, y) = (0, 0) \\ a \neq 0, 4a^3 + 27b^2 = 0, & (x, y) = (-\frac{3b}{2a}, 0) \end{cases} ,$$

it is found that when

$$4a^3 + 27b^2 = 0 \tag{7.3}$$

$(x, y) = (0, 0)$ or $(-\dfrac{3b}{2a}, 0)$ becomes a singular point (there are no others). As $(a, b) = (0, 0), (-3, 2)$ satisfy (7.3), using the same code as in Example 60, we draw its outline (Fig. 7.3). It can be seen that at the singular point $(x, y) = (0, 0)$ in the former case, no tangent line can be drawn (as it is sharp), and at the singular point $(x, y) = (1, 0)$ in the latter case, multiple tangent lines can be drawn. ∎

Next, we define the blow-up of $\mathbb{A}^2$ centered at the origin. First, we introduce a subset

$$U := \{(x, y, [x' : y']) \in \mathbb{A}^2 \times \mathbb{P}^1 \mid xy' = x'y\} \tag{7.4}$$

of $\mathbb{A}^2 \times \mathbb{P}^1$. In other words, the set U consists of two types of elements: $(x, y) \times [x : y]$ for $(x, y) \in \mathbb{A}^2 - \{(0, 0)\}$, and $(0, 0) \times [x' : y']$ for $[x' : y'] \in \mathbb{P}^1$. Furthermore, U can be written as $U_x \cup U_y$ using the set

$$U_x := \{(x, y, [x' : y']) \in U \mid x' \neq 0\}$$

$$U_y := \{(x, y, [x' : y']) \in U \mid y' \neq 0\},$$

and it becomes a manifold. In fact,

$$\phi_x : U_x \ni (u_x, u_x v_x, [1 : v_x]) \mapsto (u_x, v_x) \in \mathbb{A}^2$$

and

$$\phi_y : U_y \ni (u_y v_y, v_y, [u_y : 1]) \mapsto (u_y, v_y) \in \mathbb{A}^2$$

become a homeomorphism (1 to 1 and onto mapping that is continuous in both directions). Also, we can confirm that

$$(u_x, u_x v_x, [1 : v_x]) = (u_y v_y, v_y, [u_y : 1]) \implies u_x = u_y v_y, \; v_y = u_x v_x, \; v_x u_y = 1$$

holds, so the coordinate transformation is given by

$$\phi_x(U_x \cap U_y) \ni (u_x, v_x) \mapsto (\frac{1}{v_x}, u_x v_x) = (u_y, v_y) \in \phi_y(U_x \cap U_y)$$

$$\phi_y(U_x \cap U_y) \ni (u_y, v_y) \mapsto (u_y v_y, \frac{1}{u_y}) = (u_x, v_x) \in \phi_x(U_x \cap U_y).$$

At this time, the projection $\pi : U \to \mathbb{A}^2$, which corresponds only to the $\mathbb{A}^2$ component of the assembled U excluding the $\mathbb{P}^1$ component, gives the isomorphism

$$\pi : U - \{(0, 0)\} \times \mathbb{P}^1 \ni (x, y, [x : y]) \mapsto (x, y) \in \mathbb{A}^2 - \{(0, 0)\},$$

when excluding the origin.

Next, we consider the algebraic set $V \subseteq \mathbb{A}^d$ with $d \geq 2$. Consider the subset U' of $U \cap (V \times \mathbb{P}^{d-1})$ restricted to $\mathbb{A}^d \times \mathbb{P}^{d-1}$, where $\pi(U') = V - (0, 0)$. Here, U' does not include elements of $(0, 0) \times \mathbb{P}^{d-1}$. Therefore, it does not generally become an algebraic set. For example,

$$\{(x, y) \in \mathbb{A}^2 \mid y^2 = x^3 + x^2\} - \{(0, 0)\}$$

is not an algebraic set. However, adding $(0, 0)$ to it yields an algebraic set $(x, y) \in \{\mathbb{A}^2 \mid y^2 = x^3 + x^2\}$. In this way, for a subset U' of $\mathbb{A}^d \times \mathbb{P}^{d-1}$, the smallest algebraic set containing it is called its closure, and it is written as $\overline{U'}$. Then, the projection $\pi : U \to \mathbb{A}^d$ is called the *origin-centered blow-up* of the algebraic set V, and the manifold $\overline{U'} = \pi^{-1}(V - (0, 0))$ is called the *strict pullback* of V.

Example 66 In Example 65, when $(a, b) = (0, 0)$, it has a singularity only at the origin. If the local coordinates of U_x, U_y are (u_x, v_x), (u_y, v_y), they can be constructed as

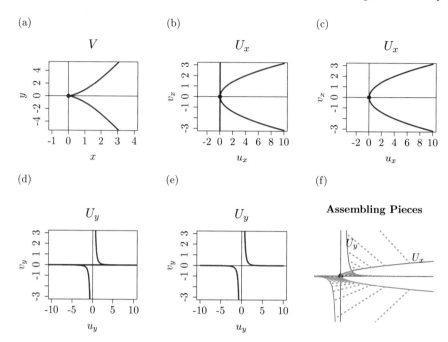

Fig. 7.4 (a) The elliptic curve $y^2 = x^3$ has a singularity at the origin. (b) It corresponds to the open set U_x. (c) Removing the blue part (curve) corresponding to the singularity in V and taking the closure (adding the origin) results in a curve without a singularity. (d) It corresponds to the open set U_y. (e) Removing the blue curve (point) corresponding to the singularity in V results in a curve without a singularity. (f) The open sets U_x, U_y correspond, except at $(0, 0)$

$$f(x, y) = f(u_x, u_x v_x) = u_x^2(v_x^2 - u_x)$$

$$f(x, y) = f(u_y v_y, v_y) = v_y^2(1 - u_y^3 v_y).$$

In the former case, $(x, y) = (0, 0) \Longleftrightarrow u_x = 0$, and in the latter case, $(x, y) = (0, 0) \Longleftrightarrow v_y = 0$, so the U' satisfying $\pi(U') = V - \{(0, 0)\}$ is $v_x^2 - u_x = 0$ when $x \neq 0$ ($u_x \neq 0$), and $1 - u_y^3 v_y = 0$ when $y \neq 0$ ($v_y \neq 0$). Taking closures of them, namely, $v_x^2 = u_x$ when $x \neq 0$, and $u_y^3 v_y = 1$ when $y \neq 0$ become the strict pull-backs, both of which are non-singular. Indeed, for $f(u_x, v_x) = v_x^2 - u_x$, $\frac{\partial f}{\partial u_x} = -1$, $\frac{\partial f}{\partial v_x} = 2v_x$, and it is impossible to make these three expressions zero simultaneously. The same is true for $f(u_y, v_y) = 1 - u_y^3 v_y$. Thus, V can be expressed in either of the local coordinates $\{(u_x, v_x) \mid v_x^2 = u_x\}, \{(u_y, v_y) \mid 1 = u_y^3 v_y\}$, and when it can be written in both, they correspond by the coordinate transformation $v_x u_y = 1$, $u_x = u_y v_y$, $v_y = u_x v_x$. In Fig. 7.4(c), (e), U_x, U_y are open sets without singularities. Also, the places where either of the local coordinates U_x, U_y can be written (except the origin) are shown in Fig. 7.4(f).   ∎

We would like to explain why generating two curves (c) (e) from the elliptic curve in Fig. 7.4(a) can be said to have resolved the singularity. First of all, (c) corresponds to $V - \{(0, 0)\}$ represented in the local coordinates (u_x, v_x) $(u_x \neq 0)$. And its pullback (the inverse image of π) is a subset of U_x. Similarly, the pullback of (e) represented in the local coordinates (u_y, v_y) $(v_y \neq 0)$ is a subset of U_y. Therefore, the strict pullback is given as a manifold as a whole. The original elliptic curve included singular points, but after finite blow-ups, when seen as a manifold, it turns out that there are no singular points in any local coordinates.

Note that if the algebraic set is not $(x, y) \in \mathbb{A}^2 \mid y^2 = x^3$, but $(x, y) \in \mathbb{A}^2 \mid y^2 = x^5$, singular points cannot be resolved with a single blow-up. For the obtained local coordinates, another blow-up is performed. In this case, the local coordinates of the manifold are further divided. In the case of $d = 2$, it is known that singular points can be resolved by repeating this process a finite number of times (Hironaka's theorem). For the general $d \geq 3$, it is necessary to apply the general blow-up introduced in the next section.

Example 67 $y^2 = x^5$ has a singular point only at the origin. If the local coordinates of U_x, U_y are (u_x, v_x), (u_y, v_y), they can each be written as

$$f(u_x, u_x v_x) = (u_x v_x)^2 - u_x^5 = u_x^2(v_x^2 - u_x^3)$$

and

$$f(u_y v_y, v_y) = v_y^2 - (u_y v_y)^5 = v_y^2(1 - u_y^5 v_y^3).$$

The term $v_x^2 - u_x^3$ in the former has a singular point at $(u_x, v_x) = (0, 0)$, and the term $1 - u_y^5 v_y^3$ in the latter is non-singular. Indeed, the term $v_x^2 - u_x^3$ can resolve the singular point if another blow-up is performed using the method in Example 66. Also, $1 - u_y^5 v_y^3$ becomes zero only when $u_y = 0$ or $v_y = 0$ when differentiating with respect to u_y, v_y, but in either case $1 - u_y^5 v_y^3$ does not become zero. ∎

Furthermore, in the case where the singular point is not the origin, as in the following example, perform a parallel shift of the coordinates and then blow up.

Example 68 In Example 65, when $(a, b) = (-3, 2)$, it has a singular point only at $(1, 0)$. If $x \mapsto x + 1$, then

$$y^2 - x^3 + 3x - 2 = y^2 - (x - 1)^2(x + 2) \mapsto y^2 - (x + 1 - 1)^2(x + 1 + 2) = y^2 - x^2(x + 3)$$

can be achieved, so consider the parallel-translated origin passing $y^2 = x^3 + 3x^2$. If the local coordinates of U_x, U_y are (u_x, v_x), (u_y, v_y), they can each be written as

$$f(u_x, u_x v_x) = (u_x v_x)^2 - u_x^3 - 3u_x^2 = u_x^2(v_x^2 - u_x - 3)$$

$$f(u_y v_y, v_y) = v_y^2 - (u_y v_y)^3 - 3(u_y v_y)^2 = v_y^2(1 - u_y^3 v_y - 3u_y^2).$$

$v_x^2 - u_x - 3 = 0$, $1 - u_y^3 v_y - 3u_y^2 = 0$ are the strict pullbacks, and both are non-singular. ∎

7.4 Hironaka's Theorem

The theory of resolving singularities in the previous section was constructed by Heisuke Hironaka in 1964.

Proposition 27 (Hironaka [4, 5]) *Let f be an analytic function from a neighborhood of the origin in $\mathbb{R}^d$ to $\mathbb{R}$, with $f(0) = 0$ and not a constant function. Then, there exists a manifold U, an open set V of $\mathbb{R}^d$ containing the origin, and an analytic map $g : U \to V$ that satisfy the following conditions:*

1. *For any compact set K of V, $g^{-1}(K)$ is a compact set of U.*
2. *Let $V_0 := \{x \in V \mid f(x) = 0\}$ and $U_0 := \{u \in U \mid f(g(u)) = 0\}$, then g gives an isomorphism[4] of $U \backslash U_0$ and $V \backslash V_0$.*
3. *For each $P \in U_0$, there exists local coordinates $(u_1, \ldots, u_d)$ of U with P as the origin, and using a multi-index $\kappa = (\kappa_1, \ldots, \kappa_d) \in \mathbb{N}^d$ and a sign $S \in \{-1, 1\}$, it can be written as*

$$f(g(u)) = S u_1^{\kappa_1} \ldots u_d^{\kappa_d}. \tag{7.5}$$

4. *The Jacobian of $x = g(u)$ can be written as an analytic function $b(u) \neq 0$, using a multi-index $h = (h_1, \ldots, h_d) \in \mathbb{N}^d$,*

$$g'(u) = b(u) u_1^{h_1} \ldots u_d^{h_d}. \tag{7.6}$$

A representation by local coordinates as in (7.5) is called a *normal crossing*. In this book, we do not prove Hironaka's theorem, but instead perform blow-ups for some specific manifolds, not algebraic sets, to find normal crossings.

Regarding Proposition 27, there are two points to note. First, in this book, we consider $K(\theta) = \mathbb{E}_X[\log \frac{p(X|\theta_*)}{p(X|\theta)}]$, $\theta \in \Theta$ as the function f. That is, it is only applied in the neighborhood of each $\theta_* \in \Theta_*$. Second, the domain of the function g is U, or it is expressed as a function of local variables. The same function symbol g is used even if the local coordinates are different, but the correspondence for each local coordinate is described.

This may be a bit late to mention,[5] in Watanabe's Bayesian theory, (so-called) singularity resolution is not used. Whether a certain $\theta_* \in \Theta_*$ is regular or not, the normal crossing form is sought. Hironaka's theorem guarantees that a normal crossing can be obtained whether it is singular or non-singular. In this sense, it may be said that there is no relationship between whether a point is regular in statistics and whether it is singular in algebraic geometry.

In the following chapters, Hironaka's theorem will be applied in the neighborhood of $\theta_* \in \Theta_*$.

---

[4] Maintains the same structure as the analytic manifold.

[5] In the field of algebraic geometry, it seems that Proposition 27 is called the singularity resolution theorem, and the process of finding the normal crossing form is called singularity resolution. However, in this book, such a description is used for the understanding of beginners.

$$y^2 - x^3 \xrightarrow{\ (x,y)=(u_1,u_1u_2)\ } -u_1^2(u_1-u_2^2) \xrightarrow{\ (u_1,u_2)=(v_1,v_1v_2)\ } -v_1^3(1-v_1v_2^2)$$

$$\Big\downarrow (x,y)=(u_1u_2,u_2)$$

$$u_2^2(1-u_1^3u_2)$$

$$\Big\downarrow (u_1,u_2)=(v_1v_2,v_2)$$

$$-v_1^2v_2^3(v_1-v_2) \xrightarrow{\ (v_1,v_2)=(w_1,w_1w_2)\ } -w_1^6w_2^3(1-w_2)$$

$$\Big\downarrow (v_1,v_2)=(w_1w_2,w_2)$$

$$w_1^2w_2^6(1-w_1)$$

$$(x,y)=\begin{cases}(u_1u_2,u_2)\\[4pt](u_1,u_1u_2)\end{cases}=\begin{cases}(v_1,v_1^2v_2)\\[4pt](v_1v_2,v_1v_2^2)\end{cases}=\begin{cases}(w_1^2w_2,\,w_1^3w_2^2)\\[4pt](w_1^2w_2,\,w_1w_2^3)\end{cases}$$

Fig. 7.5 By variable transformation, (x,y) is expressed in local coordinates (four types in this example), and the normal crossing of $y^2 - x^3$ is found

Example 69 When Hironaka's theorem is applied, Examples 66, 67, and 68 become as follows. For instance, if it is $y^2 - x^3$, the procedure is as shown in Fig. 7.5.

$f(x,y)$	$g(u)$	$g'(u)$	$f(g(u))$	S	κ	h
	$(u_1,u_1^2u_2)$	u_1^2	$-u_1^3(1-u_1u_2^2)$	-1	$(3,0)$	$(2,0)$
y^2-x^3	(u_1u_2,u_2)	u_2	$u_2^2(1-u_1^3u_2)$	1	$(0,2)$	$(0,1)$
(Example 66)	$(u_1u_2^2,u_1u_2^3)$	$u_1u_2^4$	$u_1^2u_2^6(1-u_1)$	1	$(2,6)$	$(1,4)$
	$(u_1^2u_2,u_1^3u_2^2)$	$u_1^4u_2^2$	$-u_1^6u_2^3(1-u_2)$	-1	$(6,3)$	$(4,2)$
	(u_1u_2,u_2)	u_2	$u_2^2(1-u_1^5u_2^3)$	1	$(0,2)$	$(1,0)$
y^2-x^5	$(u_1u_2,u_1u_2^2)$	$u_1u_2^2$	$u_1^2u_2^4(1-u_1^3u_2)$	1	$(2,4)$	$(1,2)$
(Example 67)	$(u_1u_2^2,u_1^2u_2^5)$	$u_1^2u_2^6$	$u_1^4u_2^{10}(1-u_1)$	1	$(4,10)$	$(2,6)$
	$(u_1^2u_2,u_1^5u_2^3)$	$u_1^6u_2^3$	$-u_1^{10}u_2(1-u_2)$	-1	$(10,1)$	$(6,3)$
	$(u_1,u_1^3u_2)$	u_1^3	$-u_1^5(1-u_1u_2^2)$	-1	$(5,0)$	$(3,0)$
$y^2-x^3-3x^2$	$(\dfrac{u_1}{\sqrt{3}},u_1u_2)$	$\dfrac{u_1}{\sqrt{3}}$	$-u_1^2(1-u_2^2+\dfrac{u_1}{3\sqrt{3}})$	-1	$(2,0)$	$(1,0)$
(Example 68)	(u_1u_2,u_2)	u_2	$u_2^2(1-u_1^3u_2-3u_1^2)$	1	$(0,2)$	$(0,1)$

■

In the first example of Example 69, $-u_1^3(1-u_1u_2^2) \approx -u_1^3$, $1-u_1u_2^2$ becomes 1 near the origin. Even if a polynomial that becomes 1 at the origin is multiplied in this way, a normal crossing can be obtained. As shown in Example 69, a normal crossing cannot be obtained by performing a variable transformation within the range of a rational map. In general, it becomes the product of a normal crossing and an analytic function that does not become 0. If the variable transformation is performed within the range of an analytic map, it becomes a normal crossing. For example, in the case of Example 66, it is sufficient to apply an analytic map that makes $u_1(1 - u_1 u_2^2)^{1/3}$ a single variable.

In the previous section, we introduced the blow-up centered at the origin for $d = 2$, but for $d = 3$, it becomes

$$\phi_x : (u_x, u_x u_y, u_x u_z, [1 : u_y : u_z])$$

$$\phi_y : (u_x u_y, u_y, u_y u_z, [u_x : 1 : u_z])$$

$$\phi_z : (u_x u_z, u_y u_z, u_z, [u_x : u_y : 1]),$$

and it is extended to the general $d > 2$. However, there may be cases where the normal crossing claimed in Hironaka's theorem cannot be obtained with the blow-up centered at the origin. From here on, we will introduce the *blow-up centered at the ideal*. The blow-up centered at the origin was

$$U = \{(0, \dots, 0)\} \times \mathbb{P}^{d-1} \cup \{(x_1, \dots, x_d, [x_1 : \cdots : x_d]) \mid (x_1, \dots, x_d) \neq (0, \dots, 0)\}$$

for the general d, but the blow-up centered at the ideal uses the ideal $I \subseteq \mathbb{R}[x]$ generated by $f_1, \dots, f_m \in \mathbb{R}[x]$, and it is set to be

$$U = V(I) \times \mathbb{P}^{m-1} \cup \{(x_1, \dots, x_d, [f_1(x_1, \dots, x_d) : \cdots :$$
$$f_m(x_1, \dots, x_d)]) \mid (x_1, \dots, x_d) \notin V(I)\}.$$

Note that

$$f_1(x_1, \dots, x_d) = 0, \dots, f_m(x_1, \dots, x_d) = 0 \Leftrightarrow (x_1, \dots, x_n) \in V(I).$$

In the blow-up centered at the ideal, it is not necessary to use all of $x_1, \dots, x_d$. The blow-up centered at the origin is equivalent to the blow-up by the ideal $(f_1(x) = x_1, \dots, f_d(x) = x_d)$. In other words, the blow-up centered at the ideal is a generalization of the blow-up centered at the origin. In Example 70, we perform a blow-up using z, x in (1), $y + \alpha_1, y$ in (2), and β_1, y in (4) as the generators of the ideal.

Example 70 For the function

$$f(x, y, z) = (xy + z)^2 + x^2 y^4,$$

we seek a normal crossing representation by local coordinates as shown in Fig. 7.6. Then, we define the mapping from each coordinate $U_i = (u_i, v_i, w_i)$, $i = 1, 2, 3, 4$ to $\mathbb{R}^3$ in

$$(1) \quad \xrightarrow{z = \alpha_1 x} \quad (2) \quad \xrightarrow{y + \alpha_1 = \beta_1 y} \quad (4) \quad \xrightarrow{\beta_1 = \gamma_1 y} \quad (6)$$

$$x = \alpha_2 z \searrow \quad (3)$$

$$y = \beta_2(y + \alpha_1)$$

$$\searrow (5) \quad y = \gamma_2 \beta_1 \quad \searrow (7)$$

(1) $\quad f = (xy + z)^2 + x^2 y^4$
(2) $\quad f = x^2\{(y + \alpha_1)^2 + y^4\}$
(3) $\quad f = z^2\{(1 + \alpha_2 y)^2 + \alpha_2^2 y^4\}$
(4) $\quad f = x^2 y^2 (\beta_1^2 + y^2)$

(5) $\quad f = x^2(y + \alpha_1)^2\{1 + \beta_2^4(y + \alpha_1)^2\}$
(6) $\quad f = x^2 y^4(1 + \gamma_1^2)$
(7) $\quad f = x^2 \gamma_2^2 \beta_1^4 (1 + \gamma_2^2)$

Fig. 7.6 The normal crossing of the function $f(x, y, z) = (xy + z)^2 + x^2 y^4$ is shown in (3) (5) (6) (7), and for this, the variables $\alpha_1, \alpha_2, \beta_1, \beta_2, \gamma_1, \gamma_2$ are introduced

$$\begin{cases} (x, y, z) = (u_1 w_1, v_1, w_1) \\ (x, y, z) = (u_2, v_2 w_2, u_2(1 - v_2)w_2) \\ (x, y, z) = (u_3, v_3, u_3 v_3(v_3 w_3 - 1)) \\ (x, y, z) = (u_4, v_4 w_4, u_4 v_4 w_4(w_4 - 1)). \end{cases}$$

However, let $\alpha_1, \alpha_2, \beta_1, \beta_2, \gamma_1, \gamma_2$ be the values defined in Fig. 7.6, and let

$$(u_1, v_1, w_1), (u_2, v_2, w_2), (u_3, v_3, w_3), (u_4, v_4, w_4)$$
$$= (\alpha_2, y, z), (x, \beta_2, y + \alpha_1), (x, y, \gamma_1), (x, \gamma_2, \beta_1),$$

respectively. For example, in the local coordinates U_1, we have $f(g(u_1, v_1, w_1)) = w_1^2(u_1 v_1 + 1)^2 + u_1^2 v_1^4$, and the Jacobian is obtained from

$$\begin{bmatrix} \frac{\partial x}{\partial u_1} & \frac{\partial x}{\partial v_1} & \frac{\partial x}{\partial w_1} \\ \frac{\partial y}{\partial u_1} & \frac{\partial y}{\partial v_1} & \frac{\partial y}{\partial w_1} \\ \frac{\partial z}{\partial u_1} & \frac{\partial z}{\partial v_1} & \frac{\partial z}{\partial w_1} \end{bmatrix} = \begin{bmatrix} w_1 & 0 & u_1 \\ 0 & 1 & 0 \\ 0 & 0 & 1 \end{bmatrix}$$

as $g'(u_1, v_1, w_1) = |w_1|$. The same calculations can be done for the others, resulting in the following.

i	U_i	$f(g(u_i, v_i, w_i))$	$g'(u_i, v_i, w_i)$	$(\kappa_1, \kappa_2, \kappa_3)$	(h_1, h_2, h_3)
1	U_1	$w_1^2\{(u_1 v_1 + 1)^2 + u_1^2 v_1^4\}$	w_1	$(0, 0, 2)$	$(0, 0, 1)$
2	U_2	$u_2^2 w_2^2(1 + v_2^4 w_2^2)$	$u_2 w_2$	$(2, 0, 2)$	$(1, 0, 1)$
3	U_3	$u_3^2 v_3^4(w_3^2 + 1)$	$u_3 v_3^2$	$(2, 4, 0)$	$(1, 2, 0)$
4	U_4	$u_4^2 v_4^2 w_4^4(1 + v_4^2)$	$u_4 v_4 w_4^2$	$(2, 2, 4)$	$(1, 1, 2)$

Also, the Jacobian $g'(u_i, v_i, w_i) \neq 0$ is the necessary and sufficient condition for the local coordinates and x, y, z are isomorphic, so it would be good to perform the pasting according to each condition of

$$g'(u_1, v_1, w_1) \neq 0 \Longleftrightarrow w_1 \neq 0 \Longleftrightarrow z \neq 0$$

$$g'(u_2, v_2, w_2) \neq 0 \Longleftrightarrow u_2 w_2 \neq 0 \Longleftrightarrow xy + z \neq 0$$

$$g'(u_3, v_3, w_3) \neq 0 \Longleftrightarrow u_3 v_3 \neq 0 \Longleftrightarrow xy \neq 0$$

$$g'(u_4, v_4, w_4) \neq 0 \Longleftrightarrow u_4 v_4 w_4 \neq 0 \Longleftrightarrow xy \neq 0.$$

Both of them can be seen to correspond to (7.5) and (7.6).   ∎

Example 71 The normal crossing representation of the function $f(x, y, z, w) = (xy + zw)^2 + (xy^2 + zw^2)^2$ by local coordinates is shown in Fig. 7.7. We try to perform the blow-up from Eq. (7.3) in two types of procedures. Although the obtained local coordinates are different, both of them are normal crossings obtained in (5) (7) (9) (10) (11). In addition, we introduce the variables $\xi_1, \xi_2, \xi_3, \alpha_1, \alpha_2, \beta_1, \beta_2, \gamma_1, \gamma_2,$ δ_1, δ_2. Also, the blow-up between (1) and (2) is symmetrical for $(x, z), (y, w)$, so we only performed the former because the same result can be obtained either by $y = \xi_1 w$ or $w = \xi_2 y$.   ∎

At the end of this section, we omit the proof but present a useful generalization for Bayesian theory by Watanabe. The specific application will be discussed in the next chapter.

Proposition 28 (*Simultaneous normal crossing[4, 5, 13]*) *Let $f_0, f_1, \ldots, f_m$ be analytic functions from a neighborhood of the origin of $\mathbb{R}^d$ to $\mathbb{R}$, where for each $i = 0, 1, \ldots, m$ we have $f_i(0) = 0$ and they are not constant functions. In this case, there exist a manifold U, an open set V in $\mathbb{R}^d$ containing the origin, and an analytic map $g : U \to V$ that satisfy the following properties:*

1. *For any compact set K in V, $g^{-1}(K)$ is a compact set in U.*
2. *g gives an isomorphism between $U \backslash U_0$ and $V \backslash V_0$, where $V_0 := \cup_{i=1}^m x \in V \mid f_i(x) = 0$ and $U_0 := \cup_{i=1}^m u \in U \mid f_i(g(u)) = 0$.*
3. *For each $P \in U_0$, there exist local coordinates $(u_1, \ldots, u_d)$ of U centered at P, multi-indices $\kappa(i) = (\kappa_1(i), \ldots, \kappa_d(i)) \in \mathbb{N}^d, i = 0, 1, \ldots, m$, analytic functions $a_i, 1 \leq i \leq m$, and a sign $S \in -1, 1$ such that we can write*

$$f_0(g(u)) = Su^{\kappa(0)}, \quad f_1(g(u)) = a_1(u)u^{\kappa(1)}, \quad \ldots, \quad f_m(g(u)) = a_m(u)u^{\kappa(m)}.$$

   Here, $u^{\kappa(i)} := u_1^{\kappa_1(i)} \ldots u_d^{\kappa_d(i)}$.
4. *The Jacobian (determinant) of $x = g(u)$ can be written as $g'(u) = b(u)u^h$, where $b(u) \neq 0$ is an analytic function that is not zero, and $h = (h_1, \ldots, h_d) \in \mathbb{N}^d$ is a multi-index.*

$$(1) \xrightarrow{\;y = \xi_1 w\;} (2) \xrightarrow{\;\xi_3 = x\xi_1 + z\;} (3)$$

$$w = \xi_2 y \searrow$$

$$(3) \xrightarrow{\;\xi_3 = \alpha_1 w\;} (4) \xrightarrow{\;\alpha_1 = \beta_1 x\;} (6) \xrightarrow{\;\beta_1 = \gamma_1 \xi_1\;} (8) \xrightarrow{\;\xi_1 - 1 = \delta_1 \gamma_1\;} (10)$$

$$w = \alpha_2 \xi_3 \searrow (5) \qquad x = \beta_2 \alpha_1 \searrow (7) \qquad \xi_1 = \gamma_2 \beta_1 \searrow (9) \qquad \searrow (11)$$

$$\gamma_1 = \delta_2(\xi_1 - 1)$$

(1) $(xy + zw)^2 + (xy^2 + zw^2)^2$	(7) $\alpha_1^2 w^4 \{1 + (\beta_2 \xi_1^2 + w - \beta_2 \xi_1)^2\}$
(2) $w^2 \{(x\xi_1 + z)^2 + w^2(x\xi_1^2 + z)^2\}$	(8) $w^4 x^2 \xi_1^2 \{\gamma_1^2 + (\xi_1 + \gamma_1 w - 1)^2\}$
(3) $w^2 \{\xi_3^2 + w^2(x\xi_1^2 + \xi_3 - x\xi_1)^2\}$	(9) $x^2 w^4 \beta_1^2 \{1 + (\beta_1 \gamma_2^2 + w - \gamma_2)^2\}$
(4) $w^4 \{\alpha_1^2 + (x\xi_1^2 + \alpha_1 w - x\xi_1)^2\}$	(10) $x^2 w^4 \{1 + \delta_1 \gamma_1\}^2 \gamma_1^2 \{1 + (\delta_1 + w)^2\}$
(5) $\alpha_2^2 \xi_3^4 \{1 + \alpha_2^2(x\xi_1^2 + \xi_3 - x\xi_1)^2\}$	(11) $w^4 x^2 \xi_1^2 (\xi_1 - 1)^2 \{\delta_2^2 + (1 + \delta_2 w)^2\}$
(6) $x^2 w^4 \{\beta_1^2 + (\xi_1^2 + \beta_1 w - \xi_1)^2\}$	

$$\beta_1 = \gamma_1(\xi_1 - 1)$$

$$(3) \xrightarrow{\;\xi_3 = \alpha_1 \xi_1\;} (4) \xrightarrow{\;\alpha_1 = \beta_1 x\;} (6) \xrightarrow{\quad} (8) \xrightarrow{\;w = \delta_1 \gamma_1\;} (10)$$

$$\xi_1 = \alpha_2 \xi_3 \searrow (5) \qquad x = \beta_2 \alpha_1 \searrow (7) \qquad \nearrow (9) \qquad \gamma_1 = \delta_2 w \searrow (11)$$

$$\xi_1 - 1 = \gamma_2 \beta_1$$

(4) $w^2 \xi_1^2 \{\alpha_1^2 + w^2(x\xi_1 + \alpha_1 - x)^2\}$	(8) $w^2 \xi_1^2 x^2 (\xi_1 - 1)^2 \{\gamma_1^2 + w^2(\gamma_1 + 1)^2\}$
(5) $w^2 \xi_3^2 \{1 + w^2(x\alpha_2^2 \xi_3 + 1 - x\alpha_2)^2\}$	(9) $x^2 w^2 (1 + \gamma_2 \beta_1)^2 \beta_1^2 \{1 + (1 + \gamma_2)^2 w^2\}$
(6) $x^2 w^2 \xi_1^2 \{\beta_1^2 + w^2(\xi_1 + \beta_1 - 1)^2\}$	(10) $x^2 \delta_1^2 \gamma_1^4 \xi_1^2 (\xi_1 - 1)^2 \{1 + \delta_1^2(\gamma_1 + 1)^2\}$
(7) $w^2 \xi_1^2 \alpha_1^2 \{1 + w^2(\beta_2 \xi_1 + 1 - \beta_2)^2\}$	(11) $w^4 x^2 \xi_1^2 (\xi_1 - 1)^2 \{\delta_2^2 + (1 + \gamma_1)^2\}$

Fig. 7.7 The normal crossing representation of the function $f(x, y, z, w) = (xy + zw)^2 + (xy^2 + zw^2)^2$ by local coordinates. We tried to perform the blow-up from Eq. (7.3) in two types of procedures. Although the obtained local coordinates are different, both are normal crossings obtained in (5) (7) (9) (10) (11)

7.5 Local Coordinates in Watanabe Bayesian Theory

As mentioned in Chap. 2, when a statistical model $p(\cdot|\theta)\theta \in \Theta$ is given, let Θ denote the set of $\theta \in \Theta$ that minimizes the Kullback-Leibler divergence with respect to the true distribution $q(\cdot)$:

$$\mathbb{E}\left[\log \frac{q(X)}{p(X|\theta)}\right].$$

In the general case without assuming regularity, Θ_* may contain multiple elements. Furthermore, since we assume finite relative variance in this book, according to

Proposition 1.3 (1), the statistical model is homogeneous, and the distribution $p(\cdot|\theta_*)$ does not depend on $\theta_* \in \Theta_*$. Hence, the function

$$K(\theta) = \mathbb{E}[\log \frac{p(X|\theta_*)}{p(X|\theta)}]$$

does not depend on $\theta_* \in \Theta_*$. In Chap. 8, we assume that $K(\cdot)$ is an analytic function. Thus, Θ_* is an analytic set. For each $\theta_* \in \Theta_*$, we shift the coordinates by the amount corresponding to θ_* and apply Proposition 27 with each of them as the origin. Then, using the local coordinates of the corresponding manifold in the neighborhood of θ_*, we express $K(\theta)$ in the form of a normal crossing.

In Chap. 5, we mentioned that we only need to remove regularity constraints on Θ contained in $B(\epsilon_n, \theta_*)$, $\theta_* \in \Theta_*$. Here, it is essential to note that the posterior distribution we want to derive in Watanabe Bayesian theory is with respect to $\Theta_m := \bigcup_{\theta_* \in \Theta_*} B(\epsilon_n, \theta_*)$, not Θ.

For each $\theta_* \in \Theta_*$, when we take θ_* as the origin, the mapping from Proposition 27 which is $g : U \to V$ can be denoted as: $g : U(\theta_*) \to V(\theta_*)$ By patching these together, it is given by: $g : U \to V$, where $U := \bigcup_{\theta_* \in \Theta_*} U(\theta_*)$ and $V := \bigcup_{\theta_* \in \Theta_*} V(\theta_*)$. In this context, rather than setting g for each θ_* with $\theta_* = 0$, a common g adjusted by θ_* is utilized. Therefore, for each $\theta_* \in \Theta_*$, we determine the normal intersection of $f(g(u) - \theta_*)$. If V is compact, from Proposition 27.1, U is also compact. This implies U can be covered by a finite union of open sets. Notably, by merging several open sets, each can include the point u such that $g(u) = 0$.

Without loss of generality, each local coordinate of the open set can be taken as a cube of size 2 centered at some element of $g^{-1}(\Theta_*)$. Furthermore, we can partition each into 2^d pieces, and adjust their signs to set each local coordinate to $[0, 1)^d$. Such variable transformations change the Jacobian. As long as it doesn't become zero within the local coordinate, it doesn't impact discussions in the next chapter. Ultimately, each of the finitely obtained open sets is denoted as U_α.

Additionally, by taking a sufficiently large n, we have $\Theta_n \subseteq V$ And U can be restricted to $g^{-1}(\Theta_n)$.

Finally, each parameter $\theta \in \Theta_m$ can generally be written in multiple local coordinates. And usually, open sets of manifolds overlap, so in the next chapter, without losing generality, we assume the following: That is, construct a C^∞ class function $\rho_\alpha : g^{-1}(\Theta_m) \to [0, 1]$ that satisfies the following three conditions:

1. $0 \le \rho_\alpha(u) \le 1$
2. $\text{supp}\, \rho_\alpha \subseteq U_\alpha$
3. $\sum_\alpha \rho_\alpha(u) = 1$

This is called a partition of unity[6]. The support $\text{supp}\, \rho_\alpha$ is defined as the smallest closed set containing points $u \in g^{-1}(\Theta_m)$ where $\rho_\alpha(u) > 0$. For example, for $u \in U_\alpha$ and

---

[6] As in Murakami [18], the partition of unity is an existing concept in manifold theory.

$$\sigma_\alpha(u) = \begin{cases} \prod_{i=1}^{d} \exp\left(-\frac{1}{1-u_i}\right), & 0 \le u_i < 1, i = 1, \ldots, d \\ 0, & \text{otherwise} \end{cases}$$

you can set $\rho_\alpha(u) = \frac{\sigma_\alpha(u)}{\sum_{\alpha'} \sigma_{\alpha'}(u)}$.

Exercises 67–74

67. As with the one-variable polynomial $\mathbb{R}[x]$ with real coefficients, ideals can be defined for the set of all integers $\mathbb{Z}$. What kind of a set is the ideal I generated by $J = 2, 3$?

68. For the elliptic curve $y^2 = x^3 + ax + b$,

   (a) Run the program from Example 60 with the settings below, and output the elliptic curve.

   ```
1  a <- 0; b <- 0; x.min <- -1; x.max <- 5     # the first one
2  a <- -3; b <- 2; x.min <- -3; x.max <- 3    # the second one
   ```

   (b) Demonstrate that the condition for including a singular point is (7.3). Determine whether each of the following is singular or non-singular:

   $$(a, b) = (-3, 3), (1, 0), (-1, 0), (0, 0), (-3, 2).$$

   Also, where are the singular points for each singular elliptic curve?

69. With regards to the Hausdorff property of topological spaces, demonstrate the following:

   (a) A metric space M is Hausdorff. [Hint] Use the triangle inequality $dist$ $(x, y) \le dist(x, z) + dist(y, z)$, $x, y, z \in M$.

   (b) Consider the set of all integers $\mathbb{Z}$ as the whole set M. Initially, only include $2n + 1$ and $2n - 1, 2n, 2n + 1$ for each $n \in \mathbb{Z}$ in $\mathcal{U}$, and generate elements of $\mathcal{U}$ to satisfy the second and third properties of a topological space. In this case, M is not Hausdorff.

70. Derive the coordinate transformation for each of the following manifolds:

   (a) For $\mathbb{P}^d = \{(U_i, \phi_i)\}_{i=0,1,\ldots,d}$, the coordinate transformation of

   $$U_i := \{[x_0 : x_1 : \cdots : x_{i-1} : 1 : x_{i+1} : \cdots : x_d] \in \mathbb{P}^d\}$$

   $$\phi_i : U_i \ni [x_0 : x_1 : \ldots : x_{i-1} : 1 : x_{i+1} : \cdots : x_d]$$
   $$\mapsto (x_0, x_1, \ldots, x_{i-1}, x_{i+1}, \ldots, x_d) \in \mathbb{A}^d$$

from $\phi_i(U_i \cap U_j)$ to $\phi_j(U_i \cap U_j)$ is given by (7.1).

(b) For the blow-up at the origin of $\mathbb{A}^2$, the coordinate transformation of (U_x, ϕ_x), (U_y, ϕ_y) is given by

$$\phi_x(U_x \cap U_y) \ni (u_x, v_x) \mapsto (\frac{1}{v_x}, u_x v_x) = (u_y, v_y) \in \phi_y(U_x \cap U_y)$$

$$\phi_y(U_x \cap U_y) \ni (u_y, v_y) \mapsto (u_y v_y, \frac{1}{u_y}) = (u_x, v_x) \in \phi_x(U_x \cap U_y).$$

71. Show that the set

$$\{(x, y) \times [x' : y'] \in \mathbb{A}^2 \times \mathbb{P}^1 \mid xy' = x'y\}$$

matches the set below.

$$\{(x, y) \times [x' : y'] \in \mathbb{A}^2 \times \mathbb{P}^1 \mid [x : y] = [x' : y'] \text{ or } (x, y) = (0, 0)\}.$$

72. In Example 69, a normal crossing is obtained using five local coordinates for $y^2 - x^3$. Construct a figure for $y^2 - x^5$ similar to Fig. 7.5.

73. In Example 70, a normal crossing is obtained for four local coordinates. Explain the operations up to obtaining the table below.

i	U_i	$f(g(u_i, v_i, w_i))$	$g'(u_i, v_i, w_i)$	$(\kappa_1, \kappa_2, \kappa_3)$	(h_1, h_2, h_3)
1	U_1	$w_1^2\{(u_1v_1 + 1)^2 + u_1^2v_1^4\}$	w_1	$(0, 0, 2)$	$(0, 0, 1)$
2	U_2	$u_2^2w_2^2(1 + v_2^4w_2^2)$	u_2w_2	$(2, 0, 2)$	$(1, 0, 1)$
3	U_3	$u_3^2v_3^4(w_3^2 + 1)$	$u_3v_3^2$	$(2, 4, 0)$	$(1, 2, 0)$
4	U_4	$u_4^2v_4^2w_4^4(1 + v_4^2)$	$u_4v_4w_4^2$	$(2, 2, 4)$	$(1, 1, 2)$

74. In Example 71, the operations to obtain the local coordinates are performed in two ways. For each of the local coordinates (5) (7) (9) (10) (11) in the first method of Fig. 7.7, calculate the Jacobian $|g'(\cdot, \cdot, \cdot, \cdot)|$.

Chapter 8
The Essence of WAIC

Based on the introductory content of algebraic geometry learned in the previous chapter, this chapter delves into the core of Watanabe's Bayesian theory. As learned in Chap. 5, the generalization to non-regular cases assumes that even if there are multiple θ_*, the range of $B(\epsilon_n, \theta_*)$ is considered. In Watanabe's Bayesian theory, the Jacobian of the variable transformation $\theta = g(u)$ is $|g'(u)|$, and the integral $\int_{[0,1]^d} [\cdot] |g'(u)| du$ is used. The integral is calculated by integrating the value of $\mathbb{E}_X [\log \frac{p(X|\theta_*)}{p(X|g(u))}] = u^\kappa$ for the same u with t, and finally integrating it with t. In other words, we find the integral of $\int_{[0,1]^d} [\cdot] \delta(t - u^\kappa) |g'(u)| du$ as a function of t using the δ function, and ignore terms that become small when $t \to 0$. The formula obtained in this way is called the state density formula in this book. The resulting posterior distribution does not converge to a normal distribution as n increases, but is expressed in a beautiful form using empirical processes. As a result, the relationship $\mathbb{E}_{X_1 \dots X_n}[G_n] = \mathbb{E}_{X_1 \dots X_n}[WAIC_n] + o(1/n)$ (Proposition 25) holds without assuming regularity. Finally, we prove that WAIC and cross-validation show almost the same value. The generalization from Chaps. 5 and 6 to 8 has the following relationships:

	Chaps. 5 and 6 Regular	Chap. 8 General
Posterior Distribution	P. 15	P. 33
G_n, T_n	P. 18, P. 19	P. 35
$\mathbb{E}_{X_1 \dots X_n}[G_n], \mathbb{E}_{X_1 \dots X_n}[T_n]$	P. 24	P. 37
$\mathbb{E}_{X_1 \dots X_n}[G_n]$ $= \mathbb{E}_{X_1 \dots X_n}[WAIC_n] + o(1/n)$	P. 25	

("P" denotes Proposition)

J. Suzuki, *WAIC and WBIC with R Stan*,
https://doi.org/10.1007/978-981-99-3838-4_8

8.1   Formula of State Density

In the following chapters, we will need to compute integrals of the form:

$$\int_{g([0,1]^d)} [\text{function of } \theta] d\theta = \int_{[0,1]^d} [\text{function of } g(u)] |g'(u)| du.$$

As pointed out in Chap. 5, the generalization without assuming regularity is only carried out for small values of u in the range of $B(\epsilon_n, \theta_*)$, that is,

$$f(g(u)) = \mathbb{E}_X[\log \frac{p(X|\theta_*)}{p(X|g(u))}].$$

Watanabe's Bayesian theory is based on the idea that when $u^\kappa = f(g(u))$ and $|g'(u)| = b(u)|u|^h$ in Hironaka's theorem, each value of $u \in [0, 1]^d$ is classified by the same value of $t = u^\kappa$. The aim was to evaluate the integral for sufficiently small values of t.

In this section, we consider δ as a function, and for $u \in [0, 1]^d$ and $t \in [0, \infty)$, we derive the specific value of

$$\delta(t - u^\kappa)|u|^h b(u) du \tag{8.1}$$

(Proposition 30, *Formula of State Density*). Note that $b(u)$ is assumed to be an analytic function with positive values as the local coordinate u moves within $[0, 1]^d$. Moreover, the δ function is defined as follows.

A function $\delta(\cdot)$ that satisfies

$$\int \delta(x)\varphi(x)dx = \varphi(0)$$

for any infinitely differentiable function $\varphi(x)$ is called a *hyper function*. Intuitively, for $a > 0$, it can be understood as the probability density of the uniform distribution in the interval $[-a, a]$ given by

$$f_a(x) = \begin{cases} 1/2a, & |x| \leq a \\ 0, & \text{otherwise} \end{cases}, \tag{8.2}$$

when a is sufficiently close to 0. Figure 8.1 shows how the uniform distribution changes with each value of a, using the following code:

```
1  delta <- function(a, j) lines(c(-a, a, a, -a, -a), c(0, 0, 1/a, 1/a,0),
2      col=j) a.seq <- seq(0.01, 0.05, 0.01)
3  plot(0, xlim=c(-0.05, 0.05), xlab="x", ylab="fa(x)",ylim=c(0, 100), type=
4      "n",
5      main="Uniform Distribution")
6  for(a in a.seq) delta(a, a*100+1)
7  legend("topleft", legend=a.seq, col=2:6, lwd=1)
```

Uniform Distributions

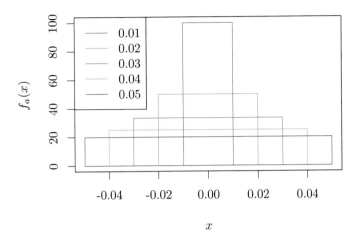

Fig. 8.1 Uniform distribution on $[-a, a]$. As a approaches 0, the values near $x = 0$ become infinitely large

We can define integration elements using generalized functions. For example, an integral using $D(x, y) = \delta(x^2 - y)$ on $\mathbb{R}^2$ is given by

$$\int_{-\infty}^{\infty} \int_{-\infty}^{\infty} f(x, y)D(x, y)dxdy = \int_{-\infty}^{\infty} f(x, x^2)dx.$$

Here, for a real-valued measurable[1] function $f(t)$ on $(0, \infty)$, the complex function $F(z)$ defined by the following is called the *Mellin transformation* of $f(t)$:

$$M : f \mapsto F, \quad F(z) := \int_0^{\infty} t^z f(t)dt, \quad z \in \mathbb{C}.$$

Although not proven in this book, the *inverse Mellin transformation*

$$M^{-1} : F \mapsto f$$

is known to uniquely exist as long as $\int_0^{\infty} t^{Re(z)} f(t)dt < \infty$ (where $Re(z)$ is the real part of z).

---

[1] For the measurability of functions, refer to 3.3.1 Random Variables.

Example 72 For $m \geq 1$ and $\lambda > 0$, let

$$f_m(t) := \begin{cases} t^{\lambda-1}(-\log t)^{m-1}, & 0 < t < 1 \\ 0, & \text{otherwise.} \end{cases}$$

The Mellin transformation of $f_m(t)$ is given by

$$M[f_m](z) = \frac{(m-1)!}{(z+\lambda)^m}. \tag{8.3}$$

Equation (8.3) is derived by using integration by parts when the real part of z is greater than[2] $-\lambda$:

$$
\begin{aligned}
M[f_m](z) &= \int_0^1 t^z t^{\lambda-1}(-\log t)^{m-1} dt \\
&= [\frac{1}{z+\lambda} t^{z+\lambda}(-\log t)^{m-1}]_0^1 + \frac{m-1}{z+\lambda} \int_0^1 t^{z+\lambda-1}(-\log t)^{m-2} dt \\
&= \frac{m-1}{z+\lambda} \cdot \int_0^1 t^z t^{\lambda-1}(-\log t)^{m-2} dt = \frac{m-1}{z+\lambda} M[f_{m-1}](z) \\
&= \frac{(m-1)!}{(z+\lambda)^{m-1}} M[f_1](z) = \frac{(m-1)!}{(z+\lambda)^{m-1}} \cdot \frac{1}{z+\lambda} = \frac{(m-1)!}{(z+\lambda)^m}.
\end{aligned}
$$

∎

Next, let $h \in \mathbb{N}^d$ and $f : [0,1]^d \to \mathbb{R}_{\geq 0}$. The *Zeta function*

$$\zeta(z) := \int_{[0,1]^d} f(u)^z u^h du , \quad z \in \mathbb{C}$$

coincides with the Mellin transformation of the density of states function on $[0,1]^d$,

$$v(t) := \int_{[0,1]^d} \delta(t - f(u)) u^h du.$$

In fact, since

$$f(u)^z = \int_0^\infty t^z \delta(t - f(u)) dt,$$

the following holds:

---

[2] when the improper integral $\lim_{\epsilon \to 0} \int_\epsilon^1 t^z t^{\lambda-1}(-\log t)^{m-1} dt$ converges.

$$\zeta(z) = \int_{[0,1]^d} f(u)^z u^h du = \int_{[0,1]^d} \int_0^\infty t^z \delta(t - f(u)) u^h dt du = \int_0^\infty t^z v(t) dt.$$

$$(8.4)$$

Example 73 Let $f(u) = u^\kappa$, $\kappa = (\kappa_1, \ldots, \kappa_d)$, and $h = (h_1, \ldots, h_d) \in \mathbb{N}^d$; the Mellin transformation of the density of states function $v(t) = \int_{[0,1]^d} \delta(t - u^\kappa) u^h du$ is given by (8.4), with $\lambda_i = (h_i + 1)/\kappa_i$; we have

$$\zeta(z) = \int_{[0,1]^d} (u^\kappa)^z u^h du = \prod_{i=1}^d \int_{0 \leq u_i \leq 1} u_i^{\kappa_i z + h_i} du_i = \prod_{i=1}^d \left[\frac{u_i^{\kappa_i z + h_i + 1}}{\kappa_i z + h_i + 1} \right]_0^1$$

$$= \frac{1}{\prod_{i=1}^d [\kappa_i (z + \lambda_i)]} = \frac{1}{\prod_{i=1}^d \kappa_i} \frac{1}{\prod_{i=1}^d (z + \lambda_i)}. \qquad (8.5)$$

■

We call the value

$$\lambda := \min_{1 \leq j \leq d} \frac{h_j + 1}{\kappa_j} > 0$$

obtained from the multi-indexes $\kappa, h \in \mathbb{N}^d$ the *real log canonical threshold*. If $\kappa_i = 0$, then $(h_i + 1)/\kappa_i = \infty$. We consider the case where at least one of $\kappa_1, \ldots, \kappa_d$ is greater than or equal to 1, that is, when λ takes a finite value. We call the number of elements m in the set

$$S := \{1 \leq j \leq d \mid \lambda = \frac{h_j + 1}{\kappa_j}\}$$

the *multiplicity*. For example, if $\kappa_j = 0$, then $j \notin S$. We divide $u = (u_a, u_b) \in [0, 1]^d$ into $u_a = (u_j)_{j \in S} \in [0, 1]^m$ and the rest $u_b = (u_j)_{j \notin S} \in [0, 1]^{d-m}$. We also define $\mu := (-\lambda \kappa_j + h_j)_{j \notin S} \in \mathbb{Q}^{d-m}$ corresponding to u_b and

$$u_b^\mu := \prod_{j \notin S} u_j^{-\lambda \kappa_j + h_j}.$$

Furthermore, let $\gamma_m := [(m-1)! \prod_{j \in S} \kappa_j]^{-1}$. Then, the following proposition can be obtained.

Proposition 29 When $m = d$, that is, when $\frac{h_j+1}{\kappa_j} = \lambda$ for $j = 1, \ldots, d$, the following holds:

$$v(t) = \int_{[0,1]^d} \delta(t - u^\kappa) u^h du = \begin{cases} \gamma_d t^{\lambda-1} (-\log t)^{d-1}, & 0 < t < 1 \\ 0, & otherwise. \end{cases} \qquad (8.6)$$

Proof By applying Examples 72, 73, and (8.4) in order, we have that for $\gamma_d = [(d-1)!\prod_{j=1}^{d}\kappa_j]^{-1}$,

$$\int_0^1 t^z \cdot \gamma_d t^{\lambda-1}(-\log t)^{d-1}dt = \frac{1}{(z+\lambda)^d} \cdot \frac{1}{\prod_{j=1}^{d}\kappa_j}$$

$$= \zeta(z) = \int_0^1 \int_{[0,1]^d} t^z \cdot \delta(t-u^\kappa)u^h dudt$$

holds. By comparing the $[\cdot]$ in $\int_0^1 t^z[\cdot]dt$, the proposition is obtained from the uniqueness of the inverse Mellin transformation. ∎

Proposition 29 assumed $m = d$, but now let's consider the general case with $m < d$, i.e., the case that includes j such that $\frac{h_j+1}{\kappa_j} > \lambda$. First, if there are s different types of λ_j such that $\lambda = \lambda_1 < \cdots < \lambda_s$, then by decomposing (8.5) into partial fractions (Exercise 76),

$$\zeta(z) = C\prod_{k=1}^{s}\frac{1}{(z+\lambda_k)^{m_k}} = \sum_{k=1}^{s}\sum_{j=1}^{m_k}\frac{c_{k,j}(j-1)!}{(z+\lambda_k)^j}$$

we find that such $C \in \mathbb{R}$ and $c_{k,j} \in \mathbb{R}$ exist for $k = 1,\ldots,s$ and $j = 1,2,\ldots,m_k$. Next, note that the linearity of the Mellin transformation $M[f+g] = M[f] + M[g] = F + G$ implies the linearity of the inverse Mellin transformation $M^{-1}[F + G] = f + g = M^{-1}[F] + M^{-1}[G]$. Then, from (8.3),

$$M^{-1}[(z+\lambda_k)^{-j}] = \frac{t^{\lambda_k-1}(-\log t)^{j-1}}{(j-1)!},$$

which implies

$$M^{-1}[\zeta(z)] = M^{-1}[\sum_{k=1}^{s}\sum_{j=1}^{m_k}\frac{c_{k,j}(j-1)!}{(z+\lambda_k)^j}] = \sum_{k=1}^{s}\sum_{j=1}^{m_k}c_{k,j}(j-1)! \cdot \frac{t^{\lambda_k-1}(-\log t)^{j-1}}{(j-1)!}$$

Therefore, we have

$$v(t) = \sum_{k=1}^{s}\sum_{j=1}^{m_k}c_{k,j}t^{\lambda_k-1}(-\log t)^{j-1} \tag{8.7}$$

Thus, to obtain the leading term $t^{\lambda-1}(-\log t)^{m-1}$ of $v(t)$ as $t \to 0$ ($\lambda = \lambda_1, m = m_1$), it suffices to investigate the pole $z = -\lambda$ of $\zeta(z)$. In this way, the following proposition is obtained.

Proposition 30 (State Density Formula) *Let $du^* := \gamma_m \delta(u_a) u_b^\mu b(u) du$; as $t \to 0$,*
the following holds:

$$\delta(t - u^\kappa) u^h b(u) du = t^{\lambda-1}(-\log t)^{m-1} du^* + o(t^{\lambda-1}(-\log t)^{m-1}).$$

Proof Refer to the appendix at the end of the chapter.

Since the term $\delta(u_a)$ is included in du^*, when integrating over $u \in [0, 1]^d$, pay attention to the fact that we should integrate over the range $u_a = 0$, i.e., $u_b \in [0, 1]^{d-m}$.

In the next section, we will examine the behavior of $\theta \in \Theta$ contained in $B(\epsilon_n, \theta_*)$, which is equivalent to investigating t values close to 0. As $t \to 0$, the value of $v(t)$ is dominated by the k that minimizes λ_k and, in the case of having the same values, by $j = m_k$.

Example 74 If the state density on $[0, 1]^3$ is

$$\delta(t - x^4 y^8 z^6) x^1 y^3 z^3 dx dy dz,$$

then with $\kappa = (4, 8, 6)$ and $h = (1, 3, 3)$, we have

$$\lambda = \min\{\frac{1+1}{4}, \frac{3+1}{8}, \frac{3+1}{6}\},$$

$\lambda = 1/2$, $m = 2$, and $\mu = -\lambda\kappa_3 + h_3 = 0$. Therefore, we obtain

$$\gamma_m = [(m-1)! \prod_{j \in S} \kappa_j]^{-1} = [1! \cdot 4 \cdot 8]^{-1} = \frac{1}{32}$$

and

$$\delta(t - x^4 y^8 z^6) x^1 y^3 z^3 dx dy dz = \frac{1}{32} t^{-1/2}(-\log t)\delta(x)\delta(y) dx dy dz + o(t^{-1/2}(-\log t)).$$

■

Suppose we apply Hironaka's theorem (Proposition 27) to an analytic function f on $\mathbb{R}^d$, and obtain the normal crossing representation (7.5) (7.6) for each local coordinate U_α. Since f is not a constant function, at least one of $\kappa_1, \ldots, \kappa_d$ must be greater than 1. In this case, for each local coordinate U_α, we can compute the real log canonical threshold $\lambda^{(\alpha)} = \min_{1 \leq i \leq d} \lambda_i^{(\alpha)}$ and the corresponding multiplicity $m^{(\alpha)}$. In the following sections, we will also need to compute the minimum value of $\lambda^{(\alpha)}$, denoted by λ, and the corresponding multiplicity m (the largest of $m^{(\alpha)}$ for which $\lambda = \lambda^{(\alpha)}$). We may simply write these as λ and m.

Example 75 For Example 70,

$$(\kappa_1, \kappa_2, \kappa_3) = (0, 0, 2), (2, 0, 2), (2, 4, 0), (2, 2, 4)$$

$$(h_1, h_2, h_3) = (0, 0, 1), (1, 0, 1), (1, 2, 0), (1, 1, 2)$$
$$\left(\frac{h_1 + 1}{\kappa_1}, \frac{h_2 + 1}{\kappa_2}, \frac{h_3 + 1}{\kappa_3}\right) = (\infty, \infty, 1), (1, \infty, 1), (1, \frac{3}{4}, \infty), (1, 1, \frac{3}{4})$$

and the values of $(\lambda^{(\alpha)}, m^{(\alpha)})$ for each U_α are

$$(\lambda^{(1)}, \lambda^{(2)}, \lambda^{(3)}, \lambda^{(4)}) = (1, 1, \frac{3}{4}, \frac{3}{4})$$

and

$$(m^{(1)}, m^{(2)}, m^{(3)}, m^{(4)}) = (1, 2, 1, 1).$$

Thus, $\lambda = 3/4$ and $m = 1$.   ∎

In this book, we do not prove it, but the values of λ and m are independent of the choice of the normal crossing representation (7.5) guaranteed by Hironaka's theorem.[3] It is known that they do not depend on the choice of local coordinates.

The state density formula will be applied in the derivation of the free energy and posterior distribution in the next section (Propositions 32, 33).

8.2   Generalization of the Posterior Distribution

The generalization to the irregular case of the posterior distribution can be done only in the neighborhood of $\theta_* \in \Theta_*$. However, in the non-regular case, there may be multiple such θ_* values. Nevertheless, the assumption of having a relatively finite variance implies that Θ_* is homogeneous (Proposition 2), i.e., $p(\cdot|\theta_*) = p(\cdot|\theta_*')$ for $\theta_*, \theta_*' \in \Theta_*$. Therefore, $K(\theta)$ becomes the same function, and the same posterior distribution is obtained for any θ_*. In other words, even if Θ_* has multiple elements, they cannot be distinguished and can be regarded as identical. In the following, without loss of generality, we arbitrarily choose an element of Θ_* and denote it as θ_*.

In this section, we investigate the function of θ

---

[3] For readers specializing in algebraic geometry, you may understand that "λ and m are birational invariants".

$$K(\theta) := \mathbb{E}_X[\log \frac{p(X|\theta_*)}{p(X|\theta)}] = \int_{\mathcal{X}} f(x, \theta)q(x)dx \qquad (8.8)$$

when the log-likelihood ratio

$$f(x, \theta) = \log \frac{p(x|\theta_*)}{p(x|\theta)}$$

has a relatively finite variance. However, since we do not assume regularity, $\theta_* \in \Theta_*$ is not unique.

First, we assume the following that also implies Assumption 1.

Assumption 4 The log-likelihood

$$f(x, \theta) = \log \frac{p(x|\theta_*)}{p(x|\theta)}$$

is a $L^2(q)$-valued analytic function.

This implies that $K(\theta) = \int_{\mathcal{X}} f(x, \theta)q(x)dx$ is also a $\mathbb{R}$-valued analytic function. In fact, when $f(x, \theta) = \sum_r a_r(x)(\theta - \theta_1)^r$, we have

$$\|a_r\|_2 = \sqrt{\int_{\mathcal{X}} a_r(x)^2 q(x)dx} \geq |\int_{\mathcal{X}} a_r(x)q(x)dx|,$$

i.e.,

$$\sum_r \|a_r\|_2 |\theta - \theta_1|^r < \infty \implies \sum_r |\int_{\mathcal{X}} a_r(x)q(x)dx||\theta - \theta_1|^r < \infty$$

holds. Therefore, Hironaka's theorem (Proposition 27) can be applied to $K(\theta)$. Additionally, from the definition of $\theta_* \in \Theta_*$, we know that the Kullback-Leibler information quantity satisfies $D(\theta) \geq D(\theta_*)$, which means that (8.8) is non-negative. Also, when we set the local coordinates $g(u) = \theta \in \Theta$, using (7.5) from Proposition 27 with $\kappa = 2k \in \mathbb{N}^d$ and $S = 1$, we obtain

$$K(g(u)) = \mathbb{E}_X[\log \frac{p(X|\theta_*)}{p(X|g(u))}] = u^{2k}, \qquad (8.9)$$

where u is the local coordinate corresponding to θ. Moreover, Assumption 4 is necessary when applying the empirical process (Proposition 11) to the log-likelihood.

For now, we will fix the discussion to a single local coordinate. Firstly, the following proposition holds.

Proposition 31 When $f(x, g(u))$ has a relatively finite variance, there exists a $L^2(q)$-valued analytic function $a(x, u)$ such that

$$f(x, g(u)) = u^k a(x, u). \tag{8.10}$$

Proof If we apply Hironaka's theorem (Proposition 27) to $K(\theta) \geq 0$, we get

$$K(g(u)) = u^{2k}.$$

From the definition, we have

$$K(g(u)) = \mathbb{E}_X[f(X, g(u))]$$

and assuming that it has a relatively finite variance, there exists a constant $C > 0$ such that

$$u^{2k} \geq C \int_X q(x) f(x, g(u))^2 dx ,$$

which leads to

$$1 \geq C \int_X q(x) \{ \frac{f(x, g(u))}{u^k} \}^2 dx. \tag{8.11}$$

Moreover, by Assumption 4, the log-likelihood $f(x, g(u))$ is an analytic function of u. So, if we denote the remainder when $f(x, g(u))$ is divided by u^κ as $b(x, u)$, we can write

$$f(x, g(u)) = u^k a(x, u) + b(x, u).$$

If $b(x, u)$ is not a zero function, then $b(x, u)/u^k$ would not be bounded as $u^k \to 0$, contradicting (8.11). Therefore, $b(x, u)$ must be identically zero. ∎

In addition, by taking the average of both sides of (8.10), we have

$$\mathbb{E}_X[a(X, u)] = u^k , \tag{8.12}$$

which holds true.

In the following, we will use the validity of (8.10) and (8.12) to generalize the posterior distribution that was derived in Chap. 5 for the regular case. For that purpose, in this section, with the observed data $x_1, \ldots, x_n \in \mathcal{X}$, we will consider the function

$$\xi_n(u) := \frac{1}{\sqrt{n}} \sum_{i=1}^{n} \{u^k - a(x_i, u)\}. \tag{8.13}$$

Between ξ_n and η_n defined in (5.1), we have the relation

$$\xi_n(u) u^k = \eta_n(g(u))$$

by (8.9), (8.10), and (8.12). Therefore, from Proposition 11, since the empirical processes $\eta_1, \ldots, \eta_n$ converge in law to the Gaussian process η, $\xi_1, \xi_2, \ldots$ are also empirical processes that converge in law to the Gaussian process ξ such that

$$\xi(u)u^k = \eta(g(u)).$$

Their mean is 0, and their covariance is given by

$$\mathbb{E}_X[\xi_n(u)\xi_n(v)] = \mathbb{E}_X[\frac{1}{n}\sum_{i=1}^n\sum_{j=1}^n\{u^k - a(X_i, u^k)\}\{v^k - a(X_j, v^k)\}]$$

$$= \frac{1}{n}\sum_{i=1}^n \mathbb{E}_X[\{u^k - a(X_i, u^k)\}\{v^k - a(X_i, v^k)\}]$$

$$= \mathbb{E}_X[a(X, u)a(X, v)] - u^k v^k. \tag{8.14}$$

The operation of the mean, which was $\mathbb{E}_{X_1 \ldots X_n}[\cdot]$ for each n in $X_1, \ldots, X_n$, will be represented as $\mathbb{E}_\xi[\cdot]$ in the case of the limit Gaussian process.

The condition of having relatively finite variance mentioned in Sect. 1.3 is essentially important, but several other assumptions are also necessary.

Assumption 5 The prior probability $\varphi(\theta)$ can be expressed as the product of a non-negative valued analytic function φ_1 and a C^∞ function φ_2 that takes positive values on Θ, i.e., $\varphi(\theta) = \varphi_1(\theta)\varphi_2(\theta)$.

For example, in cases where the prior probability $\varphi(\theta) = \varphi_1(\theta)\varphi_2(\theta) = 0$ when $K(\theta) = 0$ (such as the Jeffreys prior distribution; see Sect. 8.5), one will need to resolve the singular points of $K(\theta)$ and $\varphi_1(\theta)$ simultaneously (Proposition 28).

In the following, we will proceed with the analysis by considering $K(\theta) = K(g(u))$, i.e., $\varphi = \varphi_2$, when

$$K(\theta) = \mathbb{E}_X[\log \frac{p(X|\theta_*)}{p(X|\theta)}].$$

One could interpret the C^∞ function $\rho_\alpha(u)$ introduced in Chap. 7 as the function obtained by multiplying $\varphi_2(g(u))$ by $\rho_\alpha(u)$.

Moreover, the compact

$$\Theta = \{\theta \in \mathbb{R}^d | \pi_1(\theta) \geq 0, \ldots, \pi_m(\theta) \geq 0\}$$

can also be transformed into a local coordinate

$$g^{-1}(\Theta) := \{u \in U | \pi_1(g(u)) \geq 0, \ldots, \pi_m(g(u)) \geq 0\}$$

by simultaneous normal crossing (Proposition 28). Therefore, the following assumption is necessary:[4]

Assumption 6 $\pi_1(\theta), \ldots, \pi_m(\theta)$ are analytic functions.

In fact, in order to transform $\pi_k(\theta)$ into $\pi_k(g(u))$ for each $k = 1, \ldots, m$ according to Proposition 28, $\pi_k(\theta)$ must be an analytic function.

Then, from (8.10) (8.13), we have

$$\sum_{i=1}^{n} \log \frac{p(x_i|\theta_*)}{p(x_i|g(u))} = \sum_{i=1}^{n} f(x_i, g(u)) = \sum_{i=1}^{n} u^k a(x_i, u) = nu^{2k} - \sqrt{n}u^k \xi_n(u)$$

For the posterior distribution $\prod_{i=1}^{n} p(x_i|\theta_*) \int_{\Theta} \Omega(\theta)d\theta$

$$\Omega(\theta)d\theta := \frac{\prod_{i=1}^{n} p(x_i|\theta)}{\prod_{i=1}^{n} p(x_i|\theta_*)} \varphi(\theta)d\theta = \exp\{-nu^{2k} + \sqrt{n}u^k \xi_n(u)\}|u^h|b(u)du$$

using the properties of the δ function, we obtain

$$\Omega(\theta)d\theta = \int_0^\infty d\tau \delta(\tau - u^{2k})u^h \exp\{-n\tau + \sqrt{n\tau}\xi_n(u)\}b(u)du$$

$$= \int_0^\infty \frac{dt}{n}\delta(\frac{t}{n} - u^{2k})u^h \exp\{-t + \sqrt{t}\xi_n(u)\}b(u)du.$$

where the integration is only done with respect to dt, and $\tau := t/n$ with $d\tau = dt/n$. Moreover, by applying Proposition 30, assuming $du^* := \gamma_m\delta(u_a)u_b^h b(u)du$, we can obtain

$$\Omega(\theta)d\theta = \int_0^\infty (\frac{t}{n})^{\lambda-1}(-\log\frac{t}{n})^{m-1}\frac{dt}{n}\exp\{-t + \sqrt{t}\xi_n(u)\}du^* + o_P(n^{-\lambda})$$

$$= \frac{(\log n)^{m-1}}{n^\lambda}\int_0^\infty dt\, t^{\lambda-1}\exp\{-t + \sqrt{t}\xi_n(u)\}du^* \tag{8.15}$$

$$+\frac{1}{n^\lambda}\int_0^\infty dt\, t^{\lambda-1}(-\log t)^{m-1}\exp\{-t + \sqrt{t}\xi_n(u)\}du^* + o_P(n^{-\lambda})$$

$$= \frac{(\log n)^{m-1}}{n^\lambda}\int_0^\infty dt\, t^{\lambda-1}\exp\{-t + \sqrt{t}\xi_n(u)\}du^* + o_P(\frac{(\log n)^{m-1}}{n^\lambda}).$$

$$\tag{8.16}$$

The marginal likelihood $\int_\Theta \prod_{i=1}^{n} p(x_i|\theta)\varphi(\theta)d\theta$ is obtained by multiplying $\prod_{i=1}^{n} p(x_i|\theta_*)$ by

$$\sum_\alpha \frac{(\log n)^{m^{(\alpha)}-1}}{n^{\lambda^{(\alpha)}}} I_\alpha + o_P(\frac{(\log n)^{m-1}}{n^\lambda}), \tag{8.17}$$

---

[4] This is a theoretical assumption and not something to be aware of in practice.

where $du_\alpha^* := \gamma_{m^{(\alpha)}} \delta(u_a) u_b^{\mu_\alpha} b(u) \rho_\alpha(u) du$ and

$$I_\alpha := \int_{[0,1]^d} du_\alpha^* \int_0^\infty dt\, t^{\lambda^{(\alpha)}-1} \exp\{-t + \sqrt{t}\xi_n(u)\}. \tag{8.18}$$

The reason for attaching the subscript α to μ, λ, m is that the minimum value of $\min_{1\le i\le d} \frac{h_{\alpha,i}+1}{2k_{\alpha,i}}$ and its multiplicity are different for each local coordinate U_α. Also, the partition of a, b in (u_a, u_b) for the index set $\{1, \ldots, d\}$ also differs for each local coordinate U_α. Also, it's worth noting that in the definition of du_α^*, the function $\rho_\alpha(u)$, a partition of unity defined in Chap. 7, is involved. Furthermore, the large term $\frac{(\log n)^{m^{(\alpha)}-1}}{n^{\lambda^{(\alpha)}}}$ in the numerator and denominator of the posterior mean (defined in (8.20)) dominates the overall magnitude. That is, among the local coordinates that minimize $\lambda^{(\alpha)}$ (the minimum of these is denoted as λ), those that maximize $m^{(\alpha)}$ (the maximum of these is denoted as m) are important. Typically, there are multiple local coordinates for $\lambda^{(\alpha)}$ and $m^{(\alpha)}$. In what follows, we will refer to such λ, m as the critical real log canonical threshold and its multiplicity, and denote by A the set of local coordinates α where $m^{(\alpha)}$ is maximized among the local coordinates U_α that minimize $\lambda^{(\alpha)}$. Therefore, the real log canonocal threshold is, from (7.6), $\lambda = \min_\alpha \min_{1\le i\le d} \lambda_i^{(\alpha)}$ when the maximum pole of

$$\zeta(z) = \int_\Theta K(\theta)^z \varphi(\theta) d\theta = \sum_\alpha \int_{[0,1]^d} u^{2kz+h} b(u) du = \sum_\alpha \frac{1}{\prod_{i=1}^d (2k_i)} \frac{1}{\prod_{i=1}^d (z + \lambda_i^{(\alpha)})}$$

is $z = -\lambda$. At this point, the ratio of (8.17) to

$$\frac{(\log n)^{m-1}}{n^\lambda} \sum_{\alpha\in A} I_\alpha$$

converges to 1 as $n \to \infty$. In fact, we have

$$\frac{\sum_\alpha \frac{(\log n)^{m^{(\alpha)}-1}}{n^{\lambda^{(\alpha)}}} I_\alpha + o_P\left(\frac{(\log n)^{m-1}}{n^\lambda}\right)}{\frac{(\log n)^{m-1}}{n^\lambda} \sum_{\alpha\in A} I_\alpha} = 1 + \sum_{\alpha\notin A} \frac{(\log n)^{m^{(\alpha)}-m}}{n^{\lambda^{(\alpha)}-\lambda}} \cdot \frac{I_\alpha}{\sum_{\alpha'\in A} I_{\alpha'}}.$$

Therefore, we have

$$\log\left(\sum_{\alpha} \frac{(\log n)^{m^{(\alpha)}-1}}{n^{\lambda^{(\alpha)}}} I_{\alpha}\right)$$

$$= \log\left[\frac{(\log n)^{m-1}}{n^{\lambda}}\left(\sum_{\alpha \in A} I_{\alpha}\right)\left(1 + \sum_{\alpha \notin A} \frac{(\log n)^{m(\alpha)-m}}{n^{\lambda(\alpha)-\lambda}} \cdot \frac{I_{\alpha}}{\sum_{\alpha' \in A} I_{\alpha'}}\right)\right]$$

$$= \log\left(\frac{(\log n)^{m-1}}{n^{\lambda}} \sum_{\alpha \in A} I_{\alpha}\right) + o_P(1)$$

$$= -\lambda \log n + (m-1)\log\log n + \log\sum_{\alpha \in A} I_{\alpha} + o_P(1).$$

This leads to the following proposition.

Proposition 32 *Under the condition of a relatively finite variance, the free energy has the following asymptotic behavior:*

$$F_n = \sum_{i=1}^{n} -\log p(x_i|\theta_*) + \lambda \log n - (m-1)\log\log n - \log\left(\sum_{\alpha \in A} I_{\alpha}\right) + o_P(1).$$

$$(8.19)$$

Here, λ, m are the critical real log canonical threshold and its multiplicity, and I_{α} is the value given in (8.18).

From (8.15) and the formula of marginal likelihood $\int_{\Theta} \prod_{i=1}^{n} p(x_i|\theta)\varphi(\theta)d\theta$, we obtain the following.

Proposition 33 *Under the condition of a relatively finite variance, without assuming that the true distribution is regular with respect to the statistical model, the posterior average of the function $s : [0, \infty) \times [0, 1]^d \to \mathbb{R}$ after obtaining the samples $x_1, \ldots, x_n$ is given by*

$$\mathcal{E}[s(t, u)|x_1, \ldots, x_n] := \frac{\sum_{\alpha \in A} \int_{[0,1]^d} \int_0^{\infty} s(t, u)t^{\lambda-1}e^{-t+\sqrt{t}\xi_n(u)}dtdu_{\alpha}^*}{\sum_{\alpha \in A} \int_{[0,1]^d} \int_0^{\infty} t^{\lambda-1}e^{-t+\sqrt{t}\xi_n(u)}dtdu_{\alpha}^*}, \quad (8.20)$$

where ξ_n is the analytic function determined by $x_1, \ldots, x_n$ and is defined in (8.13).

Here, what is defined in (8.20) is defined for each local coordinate fixed by the theorem of Hironaka. Furthermore, please note that the law of large numbers convergence $\xi_n \overset{d}{\to} \xi$ implies the law of large numbers convergence of the posterior distribution (8.20).

The posterior distribution (8.20) is a generalization of Proposition 15, which assumes regularity. However, it no longer follows a normal distribution. It claims to be a sum of

$$\int_0^\infty t^{\lambda-1}e^{-t+\sqrt{t}\xi_n(u)}dt\,du_\alpha^*$$

over the critical local coordinates $\alpha \in A$. Here, for

$$\prod_{i=1}^n \frac{p(x_i|g(u))}{p(x_i|\theta_*)} = e^{-nu^{2k}+\sqrt{n}u^k\xi_n(u)}$$

it can be obtained in the same way as in the case of assuming regularity from (8.10), (8.12), (8.13), and (8.14). However, in Watanabe's Bayesian theory, Proposition 30 (the formula of state density) is applied to derive the relationship between λ, m and the posterior distribution without using the assumption of regularity.

8.3 Properties of WAIC

In this section, we first generalize the generalized loss G_n and empirical loss T_n that were introduced in Chap. 4. Here, between the parameter $\theta \in \Theta$ and the parameter (t, u), we have $\theta = g(u)$ and

$$K(\theta) = u^{2k} = \frac{t}{n}. \tag{8.21}$$

which is due to the state density $\delta\left(\frac{t}{n} - u^{2k}\right)$. Thus

$$\log \frac{p(x|\theta_*)}{p(x|g(u))} = u^k a(x, u) = \sqrt{\frac{t}{n}}a(x, u). \tag{8.22}$$

which means

$$\sum_{i=1}^n \log \frac{p(x_i|\theta_*)}{p(x_i|g(u))} = nu^{2k} - \sqrt{n}u^k\xi_n(u) = t - \sqrt{t}\xi_n(u). \tag{8.23}$$

On the other hand, applying Eq. (8.22) to Proposition 17, we obtain the following proposition for equation (5.20).

Proposition 34 *When the variance is relatively finite, the kth derivative $s^{(k)}(x, \alpha)$ of*

$$s(x, \alpha) = \sum_{k=0}^\infty \frac{1}{k!}s^{(k)}(x, 0)\alpha^k$$

defined in Eq. (5.19) becomes $O_P(n^{-k/2})$. *Therefore, the mean of the residual term* $\mathbb{E}_X[\sum_{j=k}^{\infty} \frac{1}{j!} s^{(j)}(X,0)\alpha^j]$ *and the sample mean* $\frac{1}{n}\sum_{i=1}^{n}\sum_{j=k}^{\infty}\frac{1}{j!}s^{(j)}(x_i,0)\alpha^j$ *become* $O_P(n^{-k/2})$.

Proof   This is obtained by substituting Eq. (8.22) into Proposition 33.   ∎

In the following, we derive the generalization loss G_n and empirical loss T_n under the condition of having a relatively finite variance. First, note that the posterior mean and variance of

$$-\log p(x|g(u)) = -\log p(x|\theta_*) + \sqrt{\frac{t}{n}}a(x,u) \tag{8.24}$$

(both functions of x)

$$\mathcal{E}(x) := \mathcal{E}[-\log p(x|g(u)) \mid x_1, \ldots, x_n]$$

$$\mathcal{V}(x) := \mathcal{V}[-\log p(x|g(u)) \mid x_1, \ldots, x_n]$$

are given by

$$\mathcal{E}(x) = -\log p(x|\theta_*) + \mathcal{E}\left[\sqrt{\frac{t}{n}}a(x,u) \bigg| x_1, \ldots, x_n\right] \tag{8.25}$$

$$\mathcal{V}(x) = \mathcal{V}\left[\sqrt{\frac{t}{n}}a(x,u) \bigg| x_1, \ldots, x_n\right] = \mathcal{E}\left[\frac{t}{n}a(x,u)^2 \bigg| x_1, \ldots, x_n\right] - \mathcal{E}\left[\sqrt{\frac{t}{n}}a(x,u) \bigg| x_1, \ldots, x_n\right]^2 . \tag{8.26}$$

Here, $\mathcal{E}(\cdot)$ and $\mathcal{V}(\cdot)$ denote the posterior mean and variance with respect to t, u under ξ_n, respectively.

We shall define

$$S_\lambda(a) := \int_0^\infty t^{\lambda-1} e^{-t+a\sqrt{t}} dt , \quad \lambda > 0.$$

Then, we have $S_\lambda'(a) = S_{\lambda+\frac{1}{2}}(a)$, $S_\lambda''(a) = S_{\lambda+1}(a)$, and

$$S_{\lambda+1}(a) = \int_0^\infty (-e^{-t})'(t^\lambda e^{a\sqrt{t}})dt = \int_0^\infty e^{-t}(t^\lambda e^{a\sqrt{t}})'dt$$

$$= \int_0^\infty e^{-t}\frac{a}{2\sqrt{t}}t^\lambda e^{a\sqrt{t}}dt + \int_0^\infty e^{-t}\lambda t^{\lambda-1}e^{a\sqrt{t}}dt = \frac{a}{2}S_{\lambda+\frac{1}{2}}(a) + \lambda S_\lambda(a).$$

If we substitute $a = \xi_n(u)$ into both sides and apply

$$T[\cdot] := \sum_{\alpha \in A}\int_{[0,1]^d} du_\alpha^*[\cdot]$$

to both sides, then divide by $T[S_\lambda(\xi_n(u))]$, we obtain

$$\frac{T[S_{\lambda+1}(\xi_n(u))]}{T[S_\lambda(\xi_n(u))]} = \frac{T[\lambda S_\lambda(\xi_n(u))]}{T[S_\lambda(\xi_n(u))]} + \frac{T[\frac{\xi_n(u)}{2}S_{\lambda+\frac{1}{2}}(\xi_n(u))]}{T[S_\lambda(\xi_n(u))]}.$$

On the other hand, the denominator of (8.20) can be expressed as $T[S_\lambda(\xi_n(u))]$, and we have

$$T[S_{\lambda+1}(\xi_n(u))] = \mathcal{E}[t|x_1, \ldots, x_n] \cdot T[S_\lambda(\xi_n(u))]$$

and

$$T[\xi_n(u)S_{\lambda+1/2}(\xi_n(u))] = \mathcal{E}[\sqrt{t}\xi_n(u)|x_1, \ldots, x_n] \cdot T[S_\lambda(\xi_n(u))].$$

Therefore, we can obtain

$$\mathcal{E}[t|x_1, \ldots, x_n] = \lambda + \frac{1}{2}\mathcal{E}\left[\sqrt{t}\xi_n(u)|x_1, \ldots, x_n\right]. \tag{8.27}$$

Furthermore, from (8.27), the following proposition holds for

$$G_n := \mathbb{E}_X[\mathcal{E}[-\log p(X|g(u))|x_1, \ldots, x_n]]$$

and

$$T_n := \frac{1}{n}\sum_{i=1}^{n}\mathcal{E}[-\log p(x_i|g(u))|x_1, \ldots, x_n].$$

Proposition 35 *When the variance is relatively finite, we have*

$$G_n = \mathbb{E}_X[-\log p(X|\theta_*)] + \frac{1}{n}(\lambda + \frac{1}{2}\mathcal{E}[\sqrt{t}\xi_n(u)|x_1, \ldots, x_n]) - \frac{1}{2}\mathbb{E}_X[V(X)] + o_P(\frac{1}{n})$$
$$\tag{8.28}$$

and

$$T_n = \frac{1}{n}\sum_{i=1}^{n}\{-\log p(x_i|\theta_*)\} + \frac{1}{n}(\lambda - \frac{1}{2}\mathcal{E}[\sqrt{t}\xi_n(u)|x_1, \ldots, x_n]) - \frac{1}{2}\mathbb{E}_X[V(X)] + o_P(\frac{1}{n}).$$
$$\tag{8.29}$$

Proof (5.26) and (5.27) in Proposition 18 hold even for the non-regular case. From (8.12) and (8.21), we have

$$\mathbb{E}_X[u^k a(X, u)] = u^k \cdot \mathbb{E}_X[a(X, u)] = u^{2k} = \frac{t}{n}.$$

Furthermore, from (8.27), we have

$$\mathcal{E}[\frac{t}{n}|x_1, \ldots, x_n] = \frac{1}{n}\{\lambda + \frac{1}{2}\mathcal{E}[\sqrt{t}\xi_n(u)|x_1, \ldots, x_n]\}.$$

On the other hand, applying $\mathcal{E}[\cdot|x_1, \ldots, x_n]$ to both sides of (8.23) gives

$$\sum_{i=1}^{n} \mathcal{E}(x_i) = \sum_{i=1}^{n} - \log p(x_i|\theta_*) + \mathcal{E}[t|x_1, \ldots, x_n] - \mathcal{E}[\sqrt{t}\xi_n(u)|x_1, \ldots, x_n].$$

Applying (8.27) to this, the last two terms on the right side become $\lambda - \frac{1}{2}\mathcal{E}[\sqrt{t}$ $\xi_n(u)|x_1, \ldots, x_n]$. Moreover, due to $\mathcal{V}(x) = O_p(1/n)$ and the weak law of large numbers, we have

$$\frac{1}{n}\sum_{i=1}^{n} \mathcal{V}(x_i) = \mathbb{E}_X[\mathcal{V}(X)] + o_P(1).$$

Finally, from Proposition 34, (8.28) and (8.29) hold.   ∎

Next, we consider the functional variance, converging in law ξ_n to ξ in

$$V(\xi_n) := n\mathbb{E}_X[\mathcal{V}(X)] = \mathbb{E}_X[\mathcal{V}[\sqrt{t}a(X, u)|x_1, \ldots, x_n]]$$

to define

$$V(\xi) := \mathbb{E}_X[\mathcal{V}[\sqrt{t}a(X, u)]]$$

as the functional variance. The empirical functional variance is defined as

$$V_n := \sum_{i=1}^{n} \mathcal{V}(x_i) = \sum_{i=1}^{n} \mathcal{V}[\sqrt{\frac{t}{n}}a(x_i, u)|x_1, \ldots, x_n].$$

We denote the operation of the mean with respect to the variation of the analytic function ξ as $\mathbb{E}_\xi[\cdot]$. Under these definitions, the following proposition holds, which corresponds to the generalization of Proposition 21 that holds in the regular case.

Note that $V(\xi_n)$, $V(\xi)$, and V_n are respectively the value $- \log p(X|\theta)$ when taken posterior variance with empirical process ξ_n and averaged over X, the value $- \log p(X|\theta)$ when taken posterior variance with the Gaussian process ξ and averaged over X, and the value replacing the averaging over X in $V(\xi_n)$ with the sample average over $x_1, \ldots, x_n$.

Proposition 36 *When the variance is relatively finite, we have*

$$\mathbb{E}_{X_1 \cdots X_n}\left[\mathcal{E}[\sqrt{t}\xi_n(u)|X_1, \ldots, X_n]\right] = \mathbb{E}_{X_1 \cdots X_n}[V(\xi_n)] + o(1) \qquad (8.30)$$

and

$$\mathbb{E}_\xi\left[\mathcal{E}[\sqrt{t}\xi(u)]\right] = \mathbb{E}_\xi[V(\xi)]. \tag{8.31}$$

Proof Refer to the appendix at the end of this chapter.

In the following, we call $2\nu := \mathbb{E}_\xi[\mathcal{E}[\sqrt{t}\xi(u)]] = \mathbb{E}_\xi[V(\xi)]$ as the singular fluctuation. Propositions 35 and 36 imply the following proposition.

Proposition 37 Under the condition of relatively finite variance, we have

$$\mathbb{E}_{X_1\cdots X_n}[G_n] = \mathbb{E}_X[-\log p(X|\theta_*)] + \frac{1}{n}\lambda + o(\frac{1}{n}) \tag{8.32}$$

and

$$\mathbb{E}_{X_1\cdots X_n}[T_n] = \mathbb{E}_X\left[-\log p(X|\theta_*)\right] + \frac{1}{n}(\lambda - 2\nu) + o(\frac{1}{n}). \tag{8.33}$$

Proof From (8.30) and $V(\xi_n) = n\mathbb{E}_X[\mathcal{V}(X)]$, taking average of the both sides of (8.28) and (8.29) with respect to $X_1, \ldots, X_n$, we have (8.32) and (8.33).

Proposition 38 When the variance is relatively finite, $\mathbb{E}_{X_1,\ldots,X_n}[V_n] \xrightarrow{P} 2\nu$ as $n \to \infty$.

Proof Similar to the discussion in the proof of Proposition 35,

$$\mathbb{E}_{X_1\cdots X_n}[V_n] = \mathbb{E}_{X_1\cdots X_n}[\frac{1}{n}\sum_{i=1}^n \mathcal{V}[\sqrt{t}a(X_i,u)|\xi_n]] \to \mathbb{E}_\xi\mathbb{E}_X[\mathcal{V}[\sqrt{\frac{t}{n}}a(X,u)|\xi] = \mathbb{E}_\xi[V(\xi)] = 2\nu.$$

In this, we can consider the law of large numbers

$$\frac{1}{n}\sum_{i=1}^n \mathcal{V}[\sqrt{t}a(x_i,u)|\xi] \to \mathbb{E}_X[\mathcal{V}[\sqrt{t}a(X,u)|\xi]]$$

and the convergence in distribution $\xi_n \xrightarrow{d} \xi$ to be happening simultaneously. ∎

The value of the real log canonical threshold λ is $d/2$ if it is regular. The general case will be discussed in Chap. 9.

Moreover, from Propositions 35 and 37, we have

$$\mathbb{E}_{X_1\cdots X_n}[G_n] = \mathbb{E}_{X_1\cdots X_n}[T_n + \frac{V_n}{n}] + o(\frac{1}{n}). \tag{8.34}$$

Under (8.34), we define the *Watanabe-Akaike information criterion* as

$$WAIC_n := T_n + \frac{V_n}{n}. \tag{8.35}$$

Even without assuming regularity, Proposition 25 is established.

In the following sections, when we write the posterior mean as $\mathcal{E}[\cdot]$, it means $\mathcal{E}[\cdot | x_1, \ldots, x_n]$.

8.4   Equivalence with Cross-Validation-like Methods

Here, we consider a quantity called *cross-validation* (CV)[5],

$$
CV_n := -\frac{1}{n} \sum_{i=1}^{n} \log \mathcal{E}^{-i}[p(x_i | \theta)].
\tag{8.36}
$$

In this equation, $\mathcal{E}^{-i}[\cdot]$ represents the posterior mean for $\theta \in \Theta$ under[6] $x_1, \ldots,$ $x_{i-1}, x_{i+1}, \ldots, x_n \in \mathcal{X}$.

Generally, in CV, samples are divided into several groups. For example, if there are 10 groups, you learn from 9 groups and evaluate that learning with one group. Then, by swapping the one group to be evaluated, you perform learning and evaluation a total of 10 times, and an overall evaluation value is obtained. CV is considered to be more versatile than the information criterion. When there are n samples, the CV that divides them into n groups is called *leave one out CV (LOOCV)*. The value of (8.36) is the value of LOOCV, which calculates the posterior probability with $n - 1$ samples and evaluates $-\log \mathcal{E}^{-i}[p(x_i | \theta)]$ using one sample x_i.

In the case of CV, the posterior probability is calculated based on the values of $x_1, \ldots, x_n \in \mathcal{X}$ excluding $x_i \in \mathcal{X}$, and it is evaluated at $X = x_i$, so on average, it takes the value of the generalization loss when the sample size is reduced by 1:

$$
\mathbb{E}_{X_1 \ldots X_n}[CV_n] = \mathbb{E}_{X_1 \ldots X_{n-1}}[G_{n-1}].
\tag{8.37}
$$

Furthermore, by introducing the inverse temperature $\beta > 0$, the marginal likelihood and posterior distribution are extended as follows:

$$
Z_n(\beta) := \int_{\Theta} \varphi(\theta) \prod_{i=1}^{n} p(x_i | \theta)^{\beta} d\theta
$$

and

$$
p_{\beta}(\theta | x_1, \ldots, x_n) = \frac{1}{Z_n(\beta)} \varphi(\theta) \prod_{i=1}^{n} p(x_i | \theta)^{\beta}.
$$

---

[5] The content of this section is based on the following paper [9]: Sumio Watanabe, "Asymptotic Equivalence of Bayes Cross Validation and Widely Applicable Information Criterion in Singular Learning Theory", *Journal of Machine Learning Research* Volume 11 (2010) 3571–3594.

[6] In this section, when we write $\mathcal{E}[\cdot]$, it refers to $\mathcal{E}[\cdot | x_1, \ldots, x_n]$.

So far, we have assumed $\beta = 1$ until Chap. 5, but the propositions we have shown so far hold for any $\beta > 0$. However, for the free energy, we still consider the minus logarithm of the marginal likelihood at $\beta = 1$

$$F_n = -\log Z_n(1) = -\log Z_n.$$

In that case, (8.35) can be written as

$$WAIC_n = T_n + \frac{V_n}{n}\beta \tag{8.38}$$

with

$$T_n = -\frac{1}{n}\sum_{i=1}^{n}\log \mathcal{E}_\beta[p(x_i|\theta)]$$

and

$$V_n = \sum_{i=1}^{n}\{\mathcal{E}_\beta[(\log p(x_i|\theta))^2] - (\mathcal{E}_\beta[\log p(x_i|\theta)])^2\},$$

where the posterior averages in T_n, V_n, CV_n are replaced from $p(\theta|x_1,\ldots,x_n)$ to $p_\beta(\theta|x_1,\ldots,x_n)$. That is, for $w : \Theta \to \mathbb{R}$,

$$\mathcal{E}_\beta[w(\theta)] = \int_\Theta w(\theta)p_\beta(\theta|x_1,\ldots,x_n)d\theta$$

$$\mathcal{E}_\beta^{-i}[w(\theta)] = \int_\Theta w(\theta)p_\beta(\theta|x_1,\ldots,x_{i-1},x_{i+1},\ldots,x_n)d\theta$$

$$\mathcal{V}_\beta[w(\theta)] = \mathcal{E}_\beta[\{w(\theta) - \mathcal{E}_\beta[w(\theta)]\}^2]$$

are defined.

As this section will demonstrate, when $\beta = 1$, Eqs. (8.36) and (8.38) coincide, excluding the $o_P(1/n^2)$ term.

Here, we will use the notation

$$\mathcal{T}_n(\alpha) = \frac{1}{n}\sum_{i=1}^{n}s(x_i,\alpha).$$

where, in the definition of $s(\cdot, \cdot)$ in (5.19), the posterior mean $p(\theta | x_1, \ldots, x_n)$ for $\theta \in \Theta$ under $x_1, \ldots, x_n \in \mathcal{X}$ is treated as a generalized form with $\beta > 0$ as $p_\beta(\theta | x_1, \ldots, x_n)$. Proposition 34 holds similarly even when generalized to $\beta > 0$. At this time, CV_n and $WAIC_n$ can be written as follows.

Proposition 39

$$CV_n = \frac{1}{n} \sum_{i=1}^n \log p(x_i | \theta_*) - T_n'(0) + \left(\frac{2\beta - 1}{2} \right) T_n''(0) - \left(\frac{3\beta^2 - 3\beta + 1}{6} \right) T_n'''(0) + O_P\left(\frac{1}{n^2} \right)$$
$$(8.39)$$

$$WAIC_n = \frac{1}{n} \sum_{i=1}^n \log p(x_i | \theta_*) - T_n'(0) + \left(\frac{2\beta - 1}{2} \right) T_n''(0) - \frac{1}{6} T_n'''(0) + O_P\left(\frac{1}{n^2} \right).$$
$$(8.40)$$

Proof $\mathcal{E}_\beta^{-i}[w(\theta)] = \int_\Theta w(\theta) p_\beta(\theta | x_1, \ldots, x_{i-1}, x_{i+1}, \ldots, x_n) d\theta$ can be rewritten as

$$\frac{\int_\Theta w(\theta) \{ \prod_{j \neq i} p(x_j | \theta)^\beta \} \varphi(\theta) d\theta}{\int_\Theta \{ \prod_{j \neq i} p(x_j | \theta)^\beta \} \varphi(\theta) d\theta} = \frac{\int_\Theta w(\theta) p(x_i | \theta)^{-\beta} \{ \prod_{i=1}^n p(x_i | \theta)^\beta \} \varphi(\theta) d\theta}{\int_\Theta p(x_i | \theta)^{-\beta} \{ \prod_{i=1}^n p(x_i | \theta)^\beta \} \varphi(\theta) d\theta}$$

so we can have

$$\mathcal{E}_\beta^{-i}[w(\theta)] = \frac{\mathcal{E}_\beta[w(\theta) p(x_i | \theta)^{-\beta}]}{\mathcal{E}_\beta[p(x_i | \theta)^{-\beta}]}$$
$$(8.41)$$

$$CV_n = -\frac{1}{n} \sum_{i=1}^n \log \frac{\mathcal{E}_\beta\left[\left\{ \frac{p(x_i | \theta)}{p(x_i | \theta_*)} \right\}^{1-\beta} \right]}{\mathcal{E}_\beta\left[\left\{ \frac{p(x_i | \theta)}{p(x_i | \theta_*)} \right\}^{-\beta} \right]} + \frac{1}{n} \sum_{i=1}^n \log p(x_i | \theta_*)$$

$$= T_n(-\beta) - T_n(1 - \beta) + \frac{1}{n} \sum_{i=1}^n \log p(x_i | \theta).$$
$$(8.42)$$

where we have used the fact that $s(x, \alpha) = \log \mathcal{E}_\beta\left[\left\{ \frac{p(x_i | \theta)}{p(x_i | \theta_*)} \right\}^\alpha \right]$. Then, by expanding $T_n(\alpha)$ around $\alpha = 0$ using Taylor's theorem, and setting $\alpha = -\beta$ and $\alpha = 1 - \beta$, we have

$$T_n(-\beta) = T_n(0) - \beta T_n'(0) + \frac{\beta^2}{2} T_n''(0) - \frac{\beta^3}{6} T_n'''(0) + \frac{\beta^4}{24} T_n^{(4)}(\beta^1)$$

and

$$T_n(1 - \beta) = T_n(0) + (1 - \beta) T_n'(0) + \frac{(1 - \beta)^2}{2} T_n''(0) + \frac{(1 - \beta)^3}{6} T_n'''(0) + \frac{(1 - \beta)^4}{24} T_n^{(4)}(\beta^2),$$

where β^1, β^2 exist ($|\beta^1|$, $|\beta^2| < 1 + \beta$). Therefore, from Eq. (8.42) and Proposition 34, we have

$$CV_n = \frac{1}{n}\sum_{i=1}^{n} \log p(x_i|\theta_*) - T'_n(0) + \left(\frac{2\beta-1}{2}\right)T''_n(0) - \left(\frac{3\beta^2-3\beta+1}{6}\right)T'''_n(0) + O_P(\frac{1}{n^2}),$$

which gives Eq. (8.39). Also, by substituting $T_n = \frac{1}{n}\sum_{i=1}^{n}\log p(x_i|\theta_*) - T_n(1)$, $V_n/n = T''_n(0)$ into the definition of WAIC (8.38), and expanding $T_n(1)$ using Taylor's theorem, we obtain from Proposition 34

$$WAIC_n = \frac{1}{n}\sum_{i=1}^{n}\log p(x_i|\theta_*) - T_n(1) + \beta T''_n(0) = -T'_n(0) + \frac{2\beta-1}{2}T''_n(0) - \frac{1}{6}T'''_n(0) + O_P(\frac{1}{n^2}),$$

which gives us Eq. (8.40). ∎

Proposition 39 implies, by once again applying Proposition 34, that

$$CV_n - WAIC_n = \frac{\beta - \beta^2}{2}T'''_n(0) + o_P(\frac{1}{n^{3/2}}) = O_P(\frac{1}{n^{3/2}}). \tag{8.43}$$

By the way, if we rewrite Propositions 35 and 37 for the general inverse temperature $\beta > 0$, we get

$$G_n = \mathbb{E}_X[-\log p(X|\theta_*)] + \frac{1}{n}\frac{\lambda}{\beta} + \frac{1}{2}\mathcal{E}_\beta[\sqrt{t}\xi_n(u)]) - \frac{1}{2}\mathbb{E}_X[\mathcal{V}_\beta(X)] + o_P(\frac{1}{n}) \tag{8.44}$$

$$T_n = \frac{1}{n}\sum_{i=1}^{n}\{-\log p(x_i|\theta_*)\} + \frac{1}{n}(\frac{\lambda}{\beta} - \frac{1}{2}\mathcal{E}_\beta[\sqrt{t}\xi_n(u)]) - \frac{1}{2}\mathbb{E}_X[\mathcal{V}_\beta(X)] + o_P(\frac{1}{n}) \tag{8.45}$$

$$n\mathbb{E}_{X_1...X_n}[G_n] = n\mathbb{E}_X[-\log p(X|\theta_*)] + \frac{\lambda-\nu}{\beta} + \nu + o(1) \tag{8.46}$$

$$n\mathbb{E}_{X_1...X_n}[T_n] = n\mathbb{E}[-\log p(X|\theta_*)] + \frac{\lambda-\nu}{\beta} - \nu + o(1). \tag{8.47}$$

To end this section, we implement CV and compare its values with WAIC for $\beta > 0$. Due to the use of inverse temperature $\beta > 0$, we use the Stan code `model15.stan` . The function for CV, using Eq. (8.41), is constructed as follows:

$$-\frac{1}{n}\sum_{i=1}^{n}\log \mathcal{E}_\beta^{-i}[p(x_i|\theta)] = -\frac{1}{n}\sum_{i=1}^{n}\log \frac{\mathcal{E}_\beta[p(x_i|\theta)^{1-\beta}]}{\mathcal{E}_\beta[p(x_i|\theta)^{-\beta}]}. \tag{8.48}$$

```
1   CV <- function(log_likelihood, beta)
2     - mean(log(colMeans(exp((1-beta) * log_likelihood)))
3         - log(colMeans(exp(-beta * log_likelihood)))
4       )
```

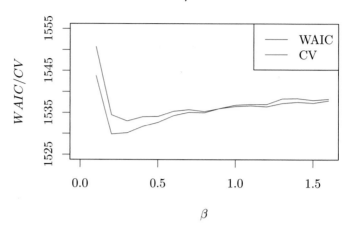

Fig. 8.2 Calculated WAIC and CV values for each $\beta > 0$ using the Boston data. As anticipated in (8.43), the difference was at most 1%

The function for `WAIC`, assuming the use of `model15.stan` constructed in Chap. 6, is extended as follows:

```
1  WAIC <- function(log_likelihood, beta)
2      T_n (log_likelihood) + beta*V_n (log_likelihood)
```

Example 76 We calculated the WAIC and CV values for each $\beta > 0$ using the Boston data (Fig. 8.2). As anticipated in (8.43), the values for both were almost identical for $\beta = 1$. Even otherwise, the difference was at most 1%.   ∎

```
1   library(rstan); library(MASS)
2   data(Boston)
3   index <- c(1, 3, 5, 6, 7, 8, 10, 11, 12, 13, 14)
4   lm <- formula(medv ~ . -medv, data=Boston)
5   df <- Boston[, index]
6   X <- model.matrix(lm, df)
7   N <- nrow(df)
8   K <- length(index)
9   Y <- df$medv
10  waic_values <- NULL
11  cv_values <- NULL
12  beta.seq <- seq(0.1, 1.6, 0.1)
13  for(beta in beta.seq){
14    data_list <- list(N=N, M=K, y=Y, x=X, beta=beta)
15    fit <- stan(file="model15.stan", data=data_list, seed=1)
16    m2 <- extract(fit)
17    waic_values <- c(waic_values, N*WAIC(m2$log_lik, beta))
18    cv_values <- c(cv_values, N*CV(m2$log_lik, beta))
19  }
```

Appendix: Proof of Proposition

Proof of Proposition 30

In the following, we seek the inverse Mellin transform of the left-hand side of the equation to be proved

$$\zeta(z) = \int_{[0,1]^d} u^{\kappa z + h} b(u) du.$$

Just like the Laplace or Fourier transforms, when seeking the inverse transform of a sum, it is sufficient to compute the inverse transforms separately and then sum them. Applying the Taylor expansion around $u_a = 0$ to $b(u_a, u_b)$, and letting

$$b(u) = b(0, u_b) + u_a^\top \nabla_a b(0, u_b) + \cdots$$

we obtain

$$
\begin{aligned}
\zeta(z) &:= \int_{[0,1]^d} u^{\kappa z + h} b(0, u_b) du + \int_{[0,1]^d} u^{\kappa z + h} u_a^\top \nabla_a b(0, u_b) du + \cdots \\
&= \left(\prod_{j \in S} \int_0^1 u_j^{\kappa_j z + h_j} du_j \right) \int_{[0,1]^{d-m}} (\prod_{j \notin S} u_j^{\kappa_j z + h_j}) b(0, u_b) du_b + \cdots \\
&= \prod_{j \in S} \frac{1}{\kappa_j z + h_j + 1} \int_{[0,1]^{d-m}} (\prod_{j \notin S} u_j^{\kappa_j z + h_j}) b(0, u_b) du_b + \cdots \\
&= \prod_{j \in S} \frac{1}{\kappa_j} \cdot \frac{1}{(z + \lambda)^m} \int_{[0,1]^{d-m}} u_b^\mu b(0, u_b) du_b + \cdots,
\end{aligned}
$$

where we set $\lambda = \min_j \frac{h_j + 1}{\kappa_j}$, and focused on the vicinity of the pole $z = -\lambda$ to observe the behavior of the inverse Mellin transform as $t \to 0$. The final $\cdots$ part does not have poles in the region $Re(z) > -\lambda$, and when it does have poles at $z = -\lambda$, their order is less than m. Therefore, it can be seen that the pole with the maximum real part of $\zeta(z)$ is $-\lambda$ and its order is m. Also, since $\mu_j > -1$, $j \notin S$, the above integral takes a finite value. Applying the inverse Mellin transform to its leading term results in

$$\prod_{j \in S} \frac{1}{\kappa_j} \cdot \frac{1}{(z + \lambda)^m} u_b^\mu b(0, u_b) \mapsto \gamma_m \delta(u_a) t^{\lambda - 1} (- \log t)^{m-1} u_b^\mu b(u_a, u_b). \tag{8.49}$$

Applying the inverse Mellin transform to terms other than the main term, the real part is less than or equal to $-\lambda$, and the order of the poles $(-\lambda)$ is less than m, so they converge to zero faster than the main term as $t \to 0$, becoming $o(t^{\lambda - 1}(- \log t)^{m-1})$.

Since $0 < t < 1$, the sum of the inverse Mellin transformed values can be written as

$$\gamma_m \delta(u_a) t^{\lambda-1} (-\log t)^{m-1} u_b^\mu b(u) du + o(t^{\lambda-1} (-\log t)^{m-1}).$$

∎

Proof of Proposition 36

Given $\beta > 0$, expressing $Z_n(\beta)/Z_{n-1}(\beta)$ in two ways yields the equality[7]

$$\frac{\int_\Theta p(x_n|\theta)^\beta \prod_{i=1}^{n-1} p(x_i|\theta)^\beta \varphi(\theta) d\theta}{Z_{n-1}(\beta)} = \left[\frac{\int_\Theta p(x_n|\theta)^{-\beta} \prod_{i=1}^{n} p(x_i|\theta)^\beta \varphi(\theta) d\theta}{Z_n(\beta)}\right]^{-1}.$$

(8.50)

Taking the logarithm of both sides of this equation, replacing the x_n on the right-hand side with $x_1, \ldots, x_n$, and taking the sum and dividing by n, and finally taking the average $\mathbb{E}_{X_1 \ldots X_n}[\cdot]$ of both sides yields

$$\mathbb{E}_{X_1 \ldots X_{n-1}} \mathbb{E}_X[\log \mathcal{E}_\beta^{-n}[\{\frac{p(X|\theta)}{p(X|\theta_*)}\}^\beta]] = -\mathbb{E}_{X_1 \ldots X_n}[\frac{1}{n} \sum_{i=1}^n \log \mathcal{E}_\beta[\{\frac{p(X|\theta)}{p(X|\theta_*)}\}^{-\beta}],$$

(8.51)

where $\mathcal{E}_\beta^{-n}[\cdot]$ is assumed to be an operation of posterior average by p_β under $x_1, \ldots, x_{n-1}$. From now on, we shall define

$$\mathcal{G}_n(\alpha) := \mathbb{E}_X[\log \mathcal{E}_\alpha[\{\frac{p(X|\theta)}{p(X|\theta_*)}\}^\alpha]] = \mathbb{E}_X[s(X, \alpha)]$$

and

$$\mathcal{T}_n(\alpha) := \frac{1}{n} \sum_{i=1}^n \log \mathcal{E}_\alpha[\{\frac{p(x_i|\theta)}{p(x_i|\theta_*)}\}^\alpha] = \frac{1}{n} \sum_{i=1}^n s(x_i, \alpha).$$

For example, we shall assume that

$$\mathcal{G}_{n-1}(\alpha) = \mathbb{E}_X[\log \mathcal{E}_\alpha^{-n}[\{\frac{p(X|\theta)}{p(X|\theta_*)}\}^\alpha].$$

In this case, (8.51) implies

---

[7] For the symbols $Z_n(\beta)$ and p_β, refer to Sect. 7.4.

$$\mathbb{E}_{X_1 \ldots X_{n-1}}[\mathcal{G}_{n-1}(\alpha)] = -\mathbb{E}_{X_1 \ldots X_n}[\mathcal{T}_n(-\alpha)]. \tag{8.52}$$

Applying the mean value theorem to both sides of (8.52) and substituting $\alpha = 1$ results in, for some $0 < \alpha^1, \alpha^2 < 1$,

$$\mathbb{E}_{X_1 \ldots X_{n-1}}[\mathcal{G}'_{n-1}(0) + \frac{1}{2}\mathcal{G}''_{n-1}(0) + \frac{1}{3!}\mathcal{G}'''_{n-1}(\alpha^1)] = \mathbb{E}_{X_1 \ldots X_n}[\mathcal{T}'_n(0) - \frac{1}{2}\mathcal{T}''_n(0) + \frac{1}{3!}\mathcal{T}'''_n(\alpha^2)]. \tag{8.53}$$

Further, we shall write

$$\begin{cases} G_n - \mathbb{E}_X[\log p(X|\theta_*)] = -\mathcal{G}_n(1) = -\mathcal{G}'_n(0) - \frac{1}{2}\mathcal{G}''_n(0) - \cdots \\ T_n - \frac{1}{n}\sum_{i=1}^n \log p(x_i|\theta_*) = -\mathcal{T}_n(1) = -\mathcal{T}'_n(0) - \frac{1}{2}\mathcal{T}''_n(0) - \cdots . \end{cases}$$

Applying the result of the proof of Proposition 35, we obtain the following. The following holds:

$$\mathcal{G}'_n(0) = \frac{1}{n}(\lambda + \frac{1}{2}\mathcal{E}[\sqrt{t}\xi_n(u)|x_1, \ldots, x_n])$$

$$\mathcal{T}'_n(0) = \frac{1}{n}(\lambda - \frac{1}{2}\mathcal{E}[\sqrt{t}\xi_n(u)|x_1, \ldots, x_n])$$

$$\mathcal{G}''_n(0) = \mathbb{E}_X[V(X)]$$

$$\mathcal{T}''_n(0) = \frac{1}{n}\sum_{i=1}^n V(x_i) = \mathbb{E}_X[V(X)] + o_P(1).$$

Moreover, since $G_n - G_{n-1} = o_P(1)$, we have

$$\mathbb{E}_{X_1 \ldots X_n}[G_n] = \mathbb{E}_{X_1 \ldots X_n}[G_{n-1}] + o(1).$$

Substituting these into (8.53), we obtain

$$\mathbb{E}_{X_1 \ldots X_n}\left[\mathcal{E}[\sqrt{t}\xi_n(u)|X_1, \ldots, X_n]\right] = n\mathbb{E}_{X_1 \ldots X_n}\mathbb{E}_X[V(X)] + o(1).$$

Defining $n\mathbb{E}_X[V(X)]$ as $V(\xi_n)$, (8.30) holds. Further, by letting $n \to \infty$, (8.31) holds, where we used the fact that $\mathcal{G}_n^{(k)}(0)$, $\mathcal{T}_n^{(k)}(0)$, $k \geq 2$ are $O_p(n^{-k/2})$ (Proposition 34), and therefore, when the average $\mathbb{E}_{X_1 \ldots X_n}[\cdot]$ is applied to them, they become $o(1)$. ∎

Exercises 75–86

75. When $\frac{h_j+1}{\kappa_j} = \lambda$, $j = 1, \ldots, d$, prove the following equation by applying Examples 72, 73, and Eq. (8.4) in order:

$$\int_0^1 t^z \cdot \gamma_d t^{\lambda-1}(-\log t)^{d-1} dt = \int_0^1 \int_{[0,1]^d} t^z \cdot \delta(t - u^\kappa) u^h du dt.$$

76. Answer the following questions about partial fraction decomposition.

    (a) Express α, β, $\gamma \in \mathbb{R}$ in terms of p, q so that they satisfy the following equation for any real number z, where $p \neq q$:

    $$\frac{1}{(z+p)^2(z+q)} = \frac{\alpha}{(z+p)^2} + \frac{\beta}{(z+p)} + \frac{\gamma}{z+q}.$$

    (b) For any real number z, we want to find $c_{k,j} \in \mathbb{R}$ that satisfy the following equation. How many conditions (linear forms) are needed for the necessary $c_{k,j}$? And does such a solution exist?, where λ_k, $k = 1, \ldots, s$ are all different:

    $$\prod_{k=1}^s \frac{1}{(z+\lambda_k)^{m_k}} = \sum_{k=1}^s \sum_{j=1}^{m_k} \frac{c_{k,j}}{(z+\lambda_k)^j}.$$

77. Show that the function in Eq. (8.2) is a probability density function for each $a > 0$. Also, display the shape of the function when $a = 10^{-3}$. [Hint] Change the upper limits of the x-axis and y-axis of the graph.

78. Why is the value of the state density function $v(t)$ as $t \to 0$ dominated by k that minimizes λ_k, and in the case where there are identical ones, k with a larger m_k?

79. About the generalization of the posterior distribution,

    (a) How was Eq. (8.15) obtained by applying the formula of state density (Proposition 30)?

    (b) Prove the following equation:

    $$\log \left(\sum_\alpha \frac{(\log n)^{m_\alpha - 1}}{n^{\lambda_\alpha}} I_\alpha \right) = -\lambda \log n + (m - 1) \log \log n + \log \left(\sum_{\alpha \in A} I_\alpha \right) + o_P(1).$$

80. For $S_\lambda(a) = \int_0^\infty t^{\lambda-1} e^{-t+a\sqrt{t}} dt$, $\lambda > 0$, prove the following:

    (a) $S'_\lambda(a) = S_{\lambda+\frac{1}{2}}(a)$

    (b) $S''_\lambda(a) = S_{\lambda+1}(a)$

    (c) $S_{\lambda+1}(a) = \frac{a}{2} S_{\lambda+\frac{1}{2}}(a) + \lambda S_\lambda(a).$

Furthermore, derive Eq. (8.27) from (c).

81. About the proof of Proposition 36,

   (a) Demonstrate Eq. (8.51) from Eq. (8.50).
   (b) What are the specific values of $G_n'(0)$, $G_n''(0)$, $T_n'(0)$, and $T_n''(0)$?

82. In the proof of Propositions 35 and 38, what kind of convergence is the following?

$$V_n = \frac{1}{n} \sum_{i=1}^{n} V\left[\sqrt{t}a(x_i, u)|x_1, \ldots, x_n\right] \rightarrow \mathbb{E}_X[V\left[\sqrt{\frac{t}{n}}a(X, u)]\right] = V(\xi)$$

$$\mathbb{E}_{X_1 \ldots X_n}[V_n] = \mathbb{E}_{X_1 \ldots X_n}[\frac{1}{n} \sum_{i=1}^{n} V[\sqrt{t}a(x_i, u)|X_1, \ldots, X_n]] \rightarrow \mathbb{E}_\xi \mathbb{E}_X[V[\sqrt{\frac{t}{n}}a(X, u)]] = \mathbb{E}_\xi[V(\xi)].$$

83. Demonstrate Eq. (8.34) from Propositions 37 and 38.
84. Prove Eqs. (8.44), (8.45), (8.46), and (8.47) for a general $\beta > 0$.
85. Derive Eq. (8.48) using Eq. (8.41). Also, why can the CV value be obtained with the following function?

```
CV <- function(log_likelihood, beta)
  -mean(log(colMeans(exp((1-beta)*log_likelihood)))
        -log(colMeans(exp(-beta*log_likelihood)))
        )
```

86. Modify the program in Example 76 to draw a graph similar to Fig. 8.2. Also, in Fig. 8.2, it can be seen that WAIC and CV show close values near $\beta = 1$. How can this be explained theoretically?

Chapter 9
WBIC and Its Application to Machine Learning

In this chapter, we examine the value of the real log canonical threshold λ. First, assuming a known value of λ, we evaluate the WBIC values for each $\beta > 0$. Then, we determine the value of λ from the WBIC values obtained for different $\beta > 0$. The value of λ is generally below $d/2$, and in the regular case, it is equal to $d/2$. We prove this result. However, there are few statistical models in which the value of λ is known in advance. In this chapter, we analytically derive the value of λ for the so-called low-rank regression, which approximates a three-layer neural network with a linear model (the proof is provided in the appendix at the end of the chapter). Furthermore, we demonstrate a method to experimentally determine the same λ and compare it with the theoretical value. Next, we introduce a method to compare the WBIC values for Gaussian mixture models and select an appropriate model. Finally, we analytically determine the value of λ when applying Jeffreys prior, known as the uninformative prior.

This chapter may contain more research-oriented content, but we encourage readers who have reached this far to thoroughly study it until the end.

9.1 Properties of WBIC

We define the Watanabe Bayesian Information Criterion (WBIC) when the inverse temperature is $\beta > 0$ as[1]

$$WBIC_n := \mathcal{E}_\beta \left[-\sum_{i=1}^{n} \log p(x_i|\theta) \right]. \tag{9.1}$$

---

[1] This section is based on the following paper [10]: Sumio Watanabe, "A Widely Applicable Bayesian Information Criterion", *Journal of Machine Learning Research* Volume 14 (2013) 867–897.

J. Suzuki, *WAIC and WBIC with R Stan*,
https://doi.org/10.1007/978-981-99-3838-4_9

On the other hand, for the generalized free energy

$$F_n(\beta) = -\log \int_\Theta \prod_{i=1}^n p(x_i|\theta)^\beta \varphi(\theta) d\theta , \quad \beta > 0 \tag{9.2}$$

($F_n(1)$ becomes the free energy F_n), we have

$$F'_n(\beta) = \mathcal{E}_\beta \left[-\sum_{i=1}^n \log p(x_i|\theta) \right] = WBIC_n$$

$$F''_n(\beta) = -\mathcal{E}_\beta \left[\{\sum_{i=1}^n \log p(x_i|\theta)\}^2 \right] + \left(\mathcal{E}_\beta [\sum_{i=1}^n \log p(x_i|\theta)] \right)^2 < 0.$$

Therefore, $WBIC_n = F'_n(\beta)$ is a monotonically decreasing function of β. Also, since $F_n(0) = 0$, applying the mean value theorem to

$$F_n = F_n(1) = \int_0^1 F'_n(\beta) d\beta,$$

we have

$$F_n = F'_n(\beta^1) = \mathcal{E}_{\beta^1}[-\sum_{i=1}^n \log p(x_i|\theta)] ,$$

where $\beta = \beta^1$ $(0 < \beta^1 < 1)$ exists. Moreover, since $F'_n(\beta)$ is a monotonically decreasing function of β, there exists an inverse temperature $\beta > 0$ such that the values of the free energy F_n and $WBIC_n$ are equal.

WBIC guarantees the performance stated in the following proposition, without assuming that the true distribution is regular with respect to the statistical model.

Proposition 40 *For $\beta_0 > 0$, the value of WBIC, $WBIC_n$, when the inverse temperature is $\beta = \beta_0 / \log n$ is such that*

$$WBIC_n = \sum_{i=1}^n -\log p(x_i|\theta_*) + \frac{\lambda \log n}{\beta_0} + U_n \sqrt{\frac{\lambda \log n}{2\beta_0}} + O_p(1) \tag{9.3}$$

exists, where $U_1, U_2, \ldots$ is a sequence of random variables that converges in law to a normal distribution with mean 0.

Proof Refer to the appendix at the end of the chapter.

9.2 Calculation of the Learning Coefficient

In this section, we consider the calculation of the real log canonical threshold λ discussed in Chaps. 7 and 8. So far, we have referred to λ as the real logarithmic eigenvalue, but in this chapter, we will call it the *learning coefficient*.

For simplicity, we shall assume that the true distribution q is realizable with respect to the statistical model, that is, there exists $\theta_* \in \Theta_*$ such that $q(\cdot) = p(\cdot|\theta_*)$. Consider the problem of finding the greatest pole $-\lambda$ and its multiplicity m for the ζ function of

$$K(\theta) = \mathbb{E}_X[\log \frac{q(X)}{p(X|\theta)}]$$

when the prior probability is φ:

$$\zeta(z) := \int_\Theta K(\theta)^z \varphi(\theta) d\theta.$$

From here on, we will call the subset of Θ

$$\text{supp}(\varphi) := \overline{\{\theta \in \Theta | \varphi(\theta) > 0\}}$$

the *support* of $\varphi(\cdot)$, where $\overline{A}$ denotes the closure of a set $A \in \mathbb{R}^d$, the smallest closed set that contains that set.

Proposition 41 *Suppose* $\text{supp}(\varphi) \cap \Theta_*$ *is not an empty set. Let λ and m be the real log canonical threshold and its multiplicity.*

1. The following quantities have positive constants.

$$\lim_{n \to \infty} \frac{n^\lambda}{(\log n)^{m-1}} \int_\Theta \exp(-nK(\theta))\varphi(\theta) d\theta \tag{9.4}$$

and

$$\lim_{t \to 0} \frac{1}{t^{\lambda-1}(-\log t)^{m-1}} \int_\Theta \delta(t - K(\theta))\varphi(\theta) d\theta. \tag{9.5}$$

2.

$$\lambda = -\lim_{n \to \infty} \frac{\log \int_\Theta \exp\{-nK(\theta)\}\varphi(\theta) d\theta}{\log n}. \tag{9.6}$$

Proof The state density formula (8.7) generally becomes an infinite sum

$$v(t) = \int_\Theta \delta(t - K(\theta))\varphi(\theta) d\theta = \sum_{k=1}^\infty \sum_{j=1}^{m_k} c_{k,j} t^{\lambda_k - 1} \{-\log t\}^{j-1}. \tag{9.7}$$

Let $M := \max_\theta K(\theta)$. Then, we have

$$\int_\Theta \exp\{-nK(\theta)\}\varphi(\theta)d\theta = \int_0^M \exp(-nt)v(t)dt = \int_0^{Mn} \exp(-\tau)v(\frac{\tau}{n}) \cdot \frac{1}{n}d\tau$$

$$= \int_0^{Mn} e^{-\tau} \sum_{k=1}^\infty \sum_{j=1}^{m_k} c_{k,j} \left(\frac{\tau}{n}\right)^{\lambda_k-1} \left\{-\log(\frac{\tau}{n})\right\}^{j-1} \frac{1}{n}d\tau$$

$$= \sum_{k=1}^\infty \sum_{j=1}^{m_k} c_{k,j} \frac{1}{n^{\lambda_k}} \int_0^{Mn} e^{-\tau}\tau^{\lambda_k-1}\{-\log\frac{\tau}{n}\}^{j-1}d\tau$$

$$= C\frac{(\log n)^{m-1}}{n^\lambda} + o(\frac{(\log n)^{m-1}}{n^\lambda}),$$

where C is a constant. Therefore, (9.4), (9.6) hold. (9.5) is obtained from (9.7).   ∎

Proposition 42 *If there exists an open set U such that $\{\theta \in U \cap \Theta_*|\varphi(\theta) > 0\}$ is not empty, $\lambda \le d/2$ holds.*

Proof Choose $\theta_* \in \Theta_*$ such that $\varphi(\theta_*) > 0$. Without loss of generality, we can assume $\theta_* = 0$. Let $\epsilon_n := n^{-1/2}$, we have

$$Z(n) := \int_\Theta \exp(-nK(\theta))\varphi(\theta)d\theta \ge \int_{\|\theta\|_2<\epsilon_n} \exp(-nK(\theta))\varphi(\theta)d\theta.$$

We will evaluate the right-hand side. From Assumption 4, since $K(\theta)$ is an analytic function, we can perform a Taylor expansion at $\theta = 0$. Also, since the true distribution is realizable, we have $K(0) = 0$, and from the assumption that $\theta_* = 0$ is an interior point of Θ_*, we get $\nabla K(\theta) = 0$. Therefore, for $K_{i,j} := \frac{\partial^2 K(\theta)}{\partial\theta_i\partial\theta_j}|_{\theta=0}, i, j = 1, \ldots, d$ and for sufficiently small $\|\theta\|_2 > 0$, we have

$$K(\theta) = \frac{1}{2}\sum_{i=1}^d \sum_{j=1}^d K_{i,j}\theta_i\theta_j + o(\|\theta\|_2^3) \tag{9.8}$$

$$Z(n) \ge \int_{\|\theta\|_2<\epsilon_n} \exp\{-\frac{n}{2}\sum_{i=1}^d \sum_{j=1}^d K_{i,j}\theta_i\theta_j - n \cdot O(\|\theta\|_2^3)\}\varphi(\theta)d\theta. \tag{9.9}$$

Note that we are not assuming regularity, so (9.8) does not necessarily become positive definite in the vicinity of $\theta = 0$. Furthermore, if we let $\theta' = \sqrt{n}\theta \in \mathbb{R}^d$, the Jacobian of the transformation is $n^{-d/2}$, and since $\|\theta\|_2 < \epsilon_n \iff \|\theta'\|_2 < 1$, the right-hand side of (9.9) becomes

$$\int_{\|\theta'\|_2<1} \exp\{-\frac{1}{2}\sum_{i=1}^d \sum_{j=1}^d K_{i,j}\theta_i'\theta_j'\}\left(1 - \frac{1}{\sqrt{n}}O(\|\theta'\|_2^3)\right)\varphi\left(\frac{\theta'}{\sqrt{n}}\right)n^{-d/2}d\theta'.$$

Moreover, from the assumption that $\varphi(0) > 0$, as $n \to \infty$, we have

$$Z(n)n^{d/2}$$

$$\geq \int_{\|\theta'\|_2 < 1} \exp\{-\frac{1}{2} \sum_{i=1}^{d} \sum_{j=1}^{d} K_{i,j}\theta_i'\theta_j'\} \left(1 - \frac{1}{\sqrt{n}} O(\|\theta'\|_2^3)\right) \varphi\left(\frac{\theta'}{\sqrt{n}}\right) d\theta'$$

$$\to \int_{\|\theta'\|_2 < 1} \exp\{-\frac{1}{2} \sum_{i=1}^{d} \sum_{j=1}^{d} K_{i,j}\theta_i'\theta_j'\} \varphi(0) d\theta' , \qquad (9.10)$$

which converges to a positive value. Furthermore, 1. in Proposition 41 means that

$$Z(n)\frac{n^{\lambda}}{(\log n)^{m-1}}$$

converges to a positive constant, but if $\lambda > d/2$, then (9.10) would converge to 0, which is a contradiction. Therefore, $\lambda \leq d/2$ holds. ∎

Proposition 43 *Let each $\theta \in \Theta$ be written as $\theta = (u, v)$, $u \in \mathbb{R}^{d_1}$, $v \in \mathbb{R}^{d_2}$. If there exists a $u_* \in \mathbb{R}^{d_1}$ that satisfies the following two conditions:*

1. *For any $v \in \mathbb{R}^{d_2}$ such that $(u_*, v) \in \Theta$, we have $K(u_*, v) = 0$ and*
2. *There exists an open set $V \subseteq \mathbb{R}^{d_2}$ such that $\varphi(u_*, v) > 0$, for all $v \in V$,*

then $\lambda \leq d_1/2$ holds.

Proof Without loss of generality, we can assume that $u_* = 0$ and $0 \in V$. In the proof of Proposition 42, we can rewrite (9.8) as

$$K(u, v) = \frac{1}{2} \sum_{i=1}^{d_1} \sum_{j=1}^{d_1} K_{i,j}u_iu_j + o(\|u\|_2^3) , \quad v \in V$$

at $u \to u_*$, and follow the same line of argument. ∎

Proposition 44 *Assume that the true distribution is regular with respect to the statistical model and is feasible. When $\varphi(\theta_*) > 0$, the following holds:*

$$\lambda = \frac{d}{2}, \quad m = 1.$$

Proof Because it is regular, $\theta_* \in \Theta_*$ is unique. Without loss of generality, we shall set $\theta_* = 0$. By applying a orthogonal transformation to θ, we can use the mean value theorem to write:

$$K(\theta) = K(0) + \theta^{\top}\nabla K(\theta)|_{\theta=0} + \frac{1}{2}\theta^{\top}\nabla^2 K(\theta^1)\theta = \frac{1}{2}\theta^{\top}J(\theta^1)\theta.$$

If we take $\epsilon > 0$ to be sufficiently small, for every θ such that $K(\theta) < \epsilon$, we can make $J(\theta^1)$ positive definite. If we denote the maximum and minimum eigenvalues of $J(\theta^1)$ when θ moves within the range of $K(\theta) \le \epsilon$ as λ_{max} and λ_{min}, respectively, we can write

$$\frac{\lambda_{min}}{2} \sum_{j=1}^{d} \theta_j^2 \le \frac{1}{2} \theta^\top J(\theta^1)\theta \le \frac{\lambda_{max}}{2} \sum_{j=1}^{d} \theta_j^2. \tag{9.11}$$

Then, define the blow-up $g : U_1 \cup \cdots \cup U_d \to \Theta$ in local coordinates $(u_1, \ldots, u_d) \in U_i$ for $i = 1, \ldots, d$ as $\theta_i = u_i$, $\theta_j = u_i u_j$ $(j \ne i)$. In this case, Eq. (9.11) can be written as follows by letting $\hat{u}_i = (\hat{u}_1, \ldots, \hat{u}_d) \in \mathbb{R}^d$ where $\hat{u}_j = u_j$ $(j \ne i)$ and $\hat{u}_i = 1$,

$$\frac{\lambda_{min}}{2} u_i^2 \{1 + \sum_{j \ne i} u_j^2\} \le \frac{1}{2} u_i^2 \hat{u}^\top J(\theta^1)\hat{u} \le \frac{\lambda_{max}}{2} u_i^2 \{1 + \sum_{j \ne i} u_j^2\}.$$

Therefore, $K(g(u))$ in the neighborhood of $u = 0$ becomes the u_i^2 multiplied by some non-zero analytic function $a(u)$. Furthermore, when $i \ne j$, we have

$$\frac{\partial \theta_j}{\partial u_k} = \begin{cases} 1, & k = i \\ u_i, & k = j \\ 0, & k \ne i, j \end{cases}, \text{ and when } i = j, \text{ we have } \frac{\partial \theta_j}{\partial u_k} = \begin{cases} 1, & k = i \\ 0, & k \ne i \end{cases}. \text{ Therefore,}$$

the determinant of the Jacobian, which has $\frac{\partial \theta_j}{\partial u_k}$ as the (j, k)-th component, is $|g'(u)| = u_i^2$ for each $i = 1, \ldots, d$. Hence, by choosing $\kappa = (0, \ldots, 0, 2, 0, \ldots, 0)$ and $h = (0, \ldots, 0, d - 1, 0, \ldots, 0)$, it follows that $\lambda = d/2$ and $m = 1$. The same argument applies to the other U_i, $i = 1, \ldots, d$.   □

Example 77 (Watanabe [12]) In this example, we consider the standard normal distribution $\mathcal{N}$ and define $\sigma(x) := e^x - 1$. We assume that the sets $\mathcal{X}$ and $\mathcal{Y}$, and the elements $x \in \mathcal{X}$ and $y \in \mathcal{Y}$, are such that the statistical model $p(y|x, \theta)$ and the true distribution $q(y)$ are given respectively by

$$p(y|x) = \frac{1}{\sqrt{2\pi}} \exp\{-\frac{(y - a\sigma(bx) - c\sigma(dx))^2}{2}\}$$

and

$$q(y) = \frac{1}{\sqrt{2\pi}} \exp\{-\frac{y^2}{2}\}.$$

We also assume that the region where the prior probability $\varphi(\theta) > 0$ is compact and that $\varphi(\theta) > 0$ for $\theta = (a, b, c, d) = (0, 0, 0, 0)$. The following equations hold:

$$\mathbb{E}_Y[\log \frac{q(Y)}{p(Y|x)}] = \frac{1}{2}\{a\sigma(bx) + c\sigma(dx)\}^2 \tag{9.12}$$

and

$$K(\theta) = \int_{\mathcal{X}} \mathbb{E}_Y[\log \frac{q(Y)}{p(Y|x)}]q(x)dx = \frac{1}{2}\int_{\mathcal{X}}\{a\sigma(bx) + c\sigma(dx)\}^2 q(x)dx,$$

where $q(x)$ is assumed to be known. Then, we define

$$h(s, \theta) := a\sigma(bs) + c\sigma(ds) = \sum_{k=1}^{\infty}\frac{s^k}{k!}(ab^k + cd^k). \qquad (9.13)$$

We can also define $p_k = ab^k + cd^k$. Then, $K(\theta) = 0$ if and only if $p_k = 0$ for all k. Next, we shall apply the local coordinates used to obtain the normal crossing (11) in the blow-up at the beginning of Example 71 in Chap. 7. We have

$$z = \xi_3 - x\xi_1 = \alpha_1 w - x\xi_1 = \beta_1 xw - x\xi_1 = \gamma_1\xi_1 xw - x\xi_1 = \delta_2\xi_1(\xi_1 - 1)xw - \xi_1 x.$$

For $u = (x, \xi_1, \delta_2, w)$, we define $g(u)$ by

$$a = x \ , \ b = \xi_1 w \ , \ c = xw(\xi_1 - 1)\xi_1\delta_2 - x\xi_1 \ , \ d = w. \qquad (9.14)$$

Then, $p_k := ab^k + cd^k$ is given by $p_1 = x\xi_1(\xi_1 - 1)\delta_2 w^2$, $p_2 = x\xi_1(\xi_1 - 1)(1 + \delta_2 w)w^2$, and for $p \geq 2$, we have

$$\begin{aligned}p_k &= x(\xi_1 w)^k + w^k\{xw(\xi_1 - 1)\xi_1\delta_2 - x\xi_1\} \\ &= x\xi_1(\xi_1 - 1)\{1 + \xi_1 + \cdots + \xi_1^{k-2} + \delta_2 w\}w^k. \qquad (9.15)\end{aligned}$$

Hence, we have

$$h(s, g(u)) = x\xi_1(\xi_1 - 1)w^2 H(s, u)$$

with

$$H(s, u) = \delta_2 s + \sum_{k=2}^{\infty}\frac{s^k}{k!}w^{k-2}\{\sum_{j=0}^{k-2}\xi_1^j + \delta_2 w\}$$

and

$$K(g(u)) = \frac{1}{2}\{x\xi_1(\xi_1 - 1)w^2\}^2\int_{\mathcal{X}} H(s, u)^2 q(s)ds.$$

Now, when we calculate the Jacobian of (9.14), the determinant becomes the product of the diagonal elements, so

$$|g'(u)| = |1 \cdot w \cdot x(\xi_1 - 1)\xi_1 w \cdot 1| = |x(\xi_1 - 1)\xi_1 w^2|. \qquad (9.16)$$

Table 9.1 The normal crossing of (5)(7)(9)(10)(11) in Example 71

	u	$g(u)$	$\dfrac{h(x, g(u))}{H(x, u)}$	$\lvert g'(u)\rvert$	Pole
(5)	$(x, \alpha_2, \xi_1, \xi_3)$	$(x, \alpha_2\xi_1\xi_3, \xi_3 - x\xi_1, \alpha_2\xi_3)$	$\alpha_2\xi_3^2$	$\lvert \alpha_2\xi_3^2\rvert$	$(-\infty, -1, -\infty, -\frac{3}{4})$
(7)	$(w, \alpha_1, \beta_2, \xi_1)$	$(\alpha_1\beta_2, w\xi_1, \alpha_1 w - \alpha_1\beta_2\xi_1 w)$	$\alpha_1 w^2$	$\lvert \alpha_1 w^2\rvert$	$(-\frac{3}{4}, -1, -\infty, -\infty)$
(9)	$(x, w, \beta_1, \gamma_2)$	$(x, w\beta_1\gamma_2, xw\beta_1 - x\beta_1\gamma_2, w)$	$xw^2\beta_1$	$\lvert xw^2\beta_1\rvert$	$(-1, -\frac{3}{4}, -1, -\infty)$
(10)	$(x, w, \gamma_1, \delta_1)$	$(x, w(1 + \delta_1\gamma_1), x(1 + \delta_1\gamma_1)(w\gamma_1 - 1), w)$	$xw^2\gamma_1$ $\cdot(1 + \delta_1\gamma_1)$	$\lvert xw^2\gamma_1$ $\cdot(1 + \delta_1\gamma_1)\rvert$	$(-1, -\frac{3}{4}, -1, -\infty)$
(11)	(x, w, δ_2, ξ_1)	$(x, \xi_1 w, xw(\xi_1 - 1)\xi_1\delta_2 - x\xi_1, w)$	$-xw^2\xi_1$ $\cdot(1 - \xi_1)$	$\lvert xw^2\xi_1$ $\cdot(1 - \xi_1)\rvert$	$(-1, -\frac{3}{4}, -\infty, -1)$

Moreover, since $\varphi(0) > 0$ and $H(s, 0) = s^2/2 \not\equiv 0$, for the local coordinates (x, w, δ_2, ξ_1), we have $\kappa = (2, 4, 0, 2)$, $h = (1, 2, 0, 1)$, so $(h + 1)/\kappa = (1, 3/4, \infty, 1)$.

Similarly, for the other normal intersections (5)(7)(9)(10) corresponding to Example 71, with local coordinates

$$(x, \alpha_2\xi_1\xi_3, \xi_3 - x\xi_1, \alpha_2\xi_3)\,,\ (\alpha_1\beta_2, w\xi_1, w\alpha_1 - \xi_1\alpha_1\beta_2, w)\,,$$

$$(x, w\beta_1\gamma_2, xw\beta_1 - x\beta_1\gamma_2, w)\,,\ (x, w(1 + \gamma_1\delta_1), x(1 + \delta_1\gamma_1)(w\gamma_1 - 1), w).$$

Similar calculations yield values of λ all greater than or equal to 3/4 (see Table 9.1). Hence, we have $\lambda = 3/4$, $m = 1$. ∎

So far, we have been mathematically determining the values of λ and m, but at the end of this section, let's consider a method for calculating them numerically from data.

Proposition 45 *Let* $\beta_{01}, \beta_{02} > 0$ *be constants, and let* $\beta_1 := \beta_{01}/\log n$ *and* $\beta_2 = \beta_{02}/\log n$. *Then,*

$$\frac{\mathcal{E}_{\beta_1}[\sum_{i=1}^{n} - \log p(x_i|\theta)] - \mathcal{E}_{\beta_2}[\sum_{i=1}^{n} - \log p(x_i|\theta)]}{1/\beta_1 - 1/\beta_2} = \lambda + O_P(1/\sqrt{\log n}) \tag{9.17}$$

holds.

Proof From Proposition 40, we have

$$\mathcal{E}_{\beta_1}[\sum_{i=1}^{n} - \log p(x_i|\theta)] = \sum_{i=1}^{n} - \log p(x_i|\theta_*) + \frac{\lambda}{\beta_1} + O_P(\sqrt{\log n})$$

Values of β and WBIC in the Boston data

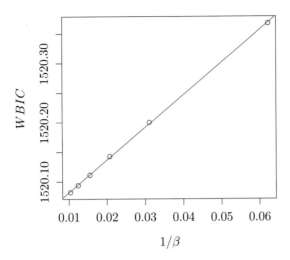

Fig. 9.1 For the Boston dataset, the intercept and slope are calculated from the pairs of $1/\beta$ and the corresponding WBIC values for $\beta = 100, 200, 300, 400, 500, 600$, and applied to the six sample points

$$\mathcal{E}_{\beta_2}[\sum_{i=1}^{n} - \log p(x_i|\theta)] = \sum_{i=1}^{n} - \log p(x_i|\theta_*) + \frac{\lambda}{\beta_2} + O_P(\sqrt{\log n}).$$

Since $1/\beta_1 - 1/\beta_2 = O_P(\log n)$, Eq. (9.17) holds. ∎

Example 78 For the Boston dataset, a regression equation was obtained from the data of pairs of $1/\beta$ and the corresponding WBIC values using the least squares method.[2] However, since the value of β becomes unstable when it is too small ($\beta < 1$) and Stan does not output the correct value when it is too large, we used the values $100, 200, 300, 400, 500, 600$ divided by $\log n$ as β, and calculated the intercept and slope when $1/\beta$ was taken on the horizontal axis and $WBIC_n$ on the vertical axis, and applied them to six sample points (Fig. 9.1). The estimated value was $\hat{\lambda} = 5.50$. Since $d = 11$, according to Proposition 44, $\lambda = d/2 = 5.5$ suggests that the true distribution was regular with respect to the statistical model. As it is a linear regression, two of the three conditions of regularity are satisfied, but it seems that the Hessian matrix $\left[\frac{\partial^2 D(q \| p(\cdot|\theta))}{\partial \theta_i \partial \theta_j} |_{\theta=\theta_*} \right] \in \mathbb{R}^{d \times d}$ was positive definite. ∎

---

[2] Execution may take a considerable amount of time depending on the PC environment.

```r
1   wbic <- function(log_likelihood) - mean(rowSums(log_likelihood))
2   library(rstan)
3   library(MASS)
4   data(Boston)
5   index <- c(1, 3, 5, 6, 7, 8, 10, 11, 12, 13, 14)
6   df <- Boston[, index]
7   lm <- formula(medv ~ . -medv, data=Boston)
8   x <- model.matrix(lm, df)
9   n <- nrow(df)
10  K <- length(index)
11  y <- df$medv
12  b.1 <- c(100,200,300,400,500,600)
13  m <- length(b.1)
14  WBIC.1 <- NULL
15  for(i in 1:m){
16    beta.1 <- b.1[i]/log(n)
17    data_list <- list(N = n, M = K, y = y, x = x, beta=beta.1)
18    fit <- stan(file = "model15.stan", data = data_list, seed = 1,iter
        =3000)
19    mm <- rstan::extract(fit)
20    wbic.1 <- wbic(mm$log_lik)
21    WBIC.1 <- c(WBIC.1,wbic.1)
22  }
23  beta.1<-b.1/log(n)
24  u<-1/beta.1
25  uu<-u-mean(u)
26  v<-WBIC.1
27  vv<-v-mean(v)
28  slope<-sum(uu*vv)/sum(uu*uu)
29  intercept<-mean(v)-slope*mean(u)
30  plot(u,v, xlab="1/beta",ylab="WBIC",main="Values of beta and WBIC in
31      the Boston data")
32  abline(a=intercept, b=slope, col="red")
```

Fig. 9.2 A three-layer neural network. The input $x_1, \ldots, x_M$ and output $y_1, \ldots, y_N$ are sandwiched by the variables $z_1, \ldots, z_H$ in the hidden layer

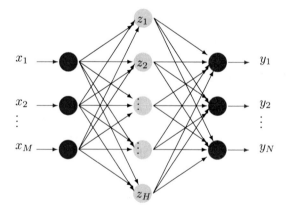

9.3 Application to Deep Learning

Let's apply the results of the previous section to a three-layer *neural network*. As shown in Fig. 9.2, suppose that the number of units in the *input layer, hidden layer,* and *output layer* are M, H, N, respectively. We use a three-layer neural network h and the probability density function $q(x)$ on $\mathbb{R}^M$ (the probability distribution of the input), and define the model as

$$p(x, y|\theta) = \frac{q(x)}{(2\pi)^{N/2}} \exp\{-\frac{1}{2}\|y - h(x, \theta)\|^2\}$$

with

$$h(x, \theta) = \sum_{k=1}^{H} a_k \sigma(x^\top b_k + c_k)$$

and

$$\sigma(t) = \tanh(t) = \frac{e^t - e^{-t}}{e^t + e^{-t}},$$

where $a_k, c_k \in \mathbb{R}, b_k \in \mathbb{R}^M, k = 1, \ldots, H$, and the true distribution $q(x)$ of $x \in \mathcal{X}$ is known. We assume that the true distribution is realizable, that is, there exists a $\theta_* \in \Theta_*$ such that $q(x, y) = p(x, y|\theta_*)$. In this case, the log-likelihood ratio is given by

$$f(x, y, \theta) = \log q(x, y) - \log p(x, y|\theta) = \frac{1}{2}[\|y - h(x, \theta)\|^2 - \|y - h(x, \theta_*)\|^2],$$

where the convergence radius of the function $\sigma(t) = \tanh(t)$ is $\pi/2$, and there exists a sequence of rational numbers $\{A_i\}$ such that

$$\sigma(t) = t - \frac{1}{3}t^3 + \frac{2}{15}t^5 - \cdots = \sum_{i=0}^{\infty} \frac{A_i}{(2i+1)!} t^{2i+1}.$$

For $x, \theta \in \mathbb{R}^d$, $\sigma(x^\top \theta)$ can be written as

$$\sum_{i=0}^{\infty} \frac{A_i}{(2i+1)!}(x^\top\theta)^{2i+1} = \sum_{i=0}^{\infty} \frac{A_i}{(2i+1)!} \sum \frac{(2i+1)!}{\alpha_1! \ldots \alpha_d!} \prod_{j=1}^{d}(x_j\theta_j)^{\alpha_j} = \sum_{\alpha} C_\alpha x^\alpha \theta^\alpha,$$

where $\sum_\alpha$ is the sum over $\alpha_1 + \cdots + \alpha_d = 2i + 1, \alpha_1, \ldots, \alpha_d \geq 0, C_\alpha$ is a constant determined by $\alpha = (\alpha_1, \ldots, \alpha_d)$, and $x^\alpha = \prod_{j=1}^d x_j^{\alpha_j}$ and $\theta^\alpha = \prod_{j=1}^d \theta_j^{\alpha_j}$. Similarly, this holds when we take $d = M + 1$, concatenate $b_k = (b_{k,1}, \ldots, b_{k,M}) \in \mathbb{R}^M$ and $c_k \in \mathbb{R}$ into $\theta_k \in \mathbb{R}^{M+1}$, and concatenate $x \in \mathbb{R}^M$ and 1 into $\in \mathbb{R}^{M+1}$ and write it again as x. That is, there exists a polynomial $h_\alpha(x)$ such that

$$h(x, \theta) = \sum_{k=1}^H a_k \sigma(x^\top b_k + c_k) = \sum_{\alpha \in \mathbb{N}^{M+1}} h_\alpha(x)\theta^\alpha ,$$

where $\alpha \in \mathbb{N}^{M+1}$ is a multi-index. Therefore, if $q(x)$ has a compact support, both $y^\top h(x, \theta)$ and $\|h(x, \theta)\|_2^2$ are bounded, and, as a result, $f(x, y, \theta)$ becomes a $L^2(q)$-valued analytic function.

When the true number of hidden units is H_*, the following facts are known about the learning coefficient λ:

1. $\lambda \leq \frac{1}{2}[H_*(M + N + 1) + \min\{(N + 1)(H - H_*), M(H - H_*), \frac{1}{3}(M(H - H_*) + 2M(N + 1))\}]$ [11]
2. When $M = N = 1$ and $H_* = 0$ [6], we have

$$\lambda = \frac{[\sqrt{H}]^2 + [\sqrt{H}] + H}{4[\sqrt{H}] + 2},$$

where $[a]$ represents the largest integer that does not exceed $a > 0$.

In the case where $\sigma(x) = x$ and $c_k = 0$, if we denote $A \in \mathbb{R}^{H \times M}, B \in \mathbb{R}^{N \times H}$, and their true values as A_*, B_*, we can express the relationship as

$$y - B_* A_* x \sim N(0, I_N).$$

This is called a *reduced-rank regression*. Reduced-rank regression is a type of regression that assumes that the rank of the linear transformation of inputs and outputs is smaller to the dimensions of the inputs and outputs. The rank of the linear transformation corresponds to the number of hidden variables. There are few examples where the exact value of the learning coefficient, rather than its upper or lower bounds, has been analytically determined. In this book, we provide a proof for the following proposition in the appendix. It is a derivation that can be understood with the knowledge gained so far, and it is recommended to try it out.

Proposition 46 (Aoyagi [7]) *In reduced-rank regression, let $\varphi(\theta_*) > 0$ be the value of the prior distribution at the true parameters, and let the rank of $B_* A_*$ be $r := H_*$. When $r \leq \min\{H, M, N\}$, the following holds.*

Case	$\begin{array}{c} M+r \\ \vdots \\ N+H \end{array}$	$\begin{array}{c} N+r \\ \vdots \\ M+H \end{array}$	$\begin{array}{c} H+r \\ \vdots \\ M+N \end{array}$	$\begin{array}{c} M+H \\ + \\ N+r \end{array}$	m	λ
1a	$\leq$	$\leq$	$\leq$	Even	1	$-(H+r-M-N)^2/8 + MN/2$
1b				Odd	2	$-(H+r-M-N)^2/8 + MN/2 + 1/8$
2	$>$				1	$(HN-Hr+Mr)/2$
3		$>$			1	$(HM-Hr+Nr)/2$
4			$>$		1	$MN/2$

Proof Refer to the appendix at the end of this chapter.

The mathematical derivation is sufficient, but we shall also try to calculate it numerically using the method from the previous section.

It is sufficient to calculate the WBIC value for each $\beta > 0$. However, to do that, we need to set the prior probability of the matrix BA.

Here, we modified the implementation of reduced-rank regression by B. Files [3] using Stan, and attempted to calculate the WBIC value. First, assume that $X \in \mathbb{R}^{n \times M}$ and $Y \in \mathbb{R}^{n \times N}$ are observed. By setting $A \in \mathbb{R}^{H \times M}$ and $B = \mathbb{R}^{N \times H}$, and assuming $e \sim N(0, \tau^2 I_n)$, we model as follows:

$$Y = X A^\top B^\top + e.$$

We want to calculate the posterior distribution of $A^\top B^\top \in \mathbb{R}^{M \times N}$. To do this, we need to set the prior distribution of $A^\top B^\top$. First, to allow the distribution of each component of $A^\top$ to vary row by row, set $A_{i,j} \sim N(0, \lambda_j)$ and $\lambda_j \sim N_+(0, 2)$. Here, $\sim N_+(0, 1)$ denotes the distribution of the absolute value of a random variable following the standard normal distribution. Let $\hat{A}_{i,j} \sim N(0, 1)$, $\Lambda = \mathrm{diag}(\lambda_1, \ldots, \lambda_M)$ (diagonal matrix), and $A^\top = \Lambda \hat{A}^T$. Then, decompose $\hat{A}^\top$ into the product of a lower triangular matrix $L \in \mathbb{R}^{M \times H}$ with non-negative diagonal elements and an orthogonal matrix $Q \in \mathbb{R}^{H \times H}$, i.e., $\hat{A}^\top = LQ$. Since Q is an orthogonal matrix, the distributions of $B^\top$ and $\hat{B}^\top := QB^\top$ are the same. We assume $\hat{B}_{ij} \sim N(0, 1)$

$$A^\top B^\top = \Lambda A^\top B^\top = \Lambda L Q B^\top = \Lambda L \hat{B}^\top,$$

where the probability that each diagonal element $L_{i,i}$ of L is $x > 0$ is proportional to

$$x^{k-i} \exp\{-\frac{x^2}{2}\}$$

(Leung-Drton's prior distribution [2]), and the off-diagonal elements follow $N(0, 1)$ (each component is independent).

On the other hand, for $e \sim N(0, \tau^2 I_n)$, the prior distribution of τ is assumed to be $\tau \sim IG(\nu, \sigma)$ with $\nu \sim G(2, 0.1)$ and $\sigma \sim Cauchy(0, 1)$, where the probability density functions of the Gamma distribution $u \sim G(\alpha, \beta)$ and the inverse Gamma distribution $v \sim IG(\nu, \sigma)$ are given by

$$f_G(u|\alpha, \beta) = \frac{u^{\alpha-1}e^{-\beta u}\beta^{\alpha}}{\Gamma(\alpha)}, \quad f_{IG}(v|\nu, \sigma) = \frac{2}{\Gamma(\frac{\nu}{2})}(\frac{\nu\sigma^2}{2})^{\frac{\nu}{2}}v^{-\nu-1}\exp(-\frac{\nu\sigma^2}{2v^2})$$

using the Gamma function Γ. Then, when $\tilde{\nu} = \nu + N$ and $\tilde{\sigma}^2 = (\sigma^2 + z^\top z)/(\nu + N)$ are set, from $\int_0^\infty f_{IG}(\tau|\tilde{\nu}, \tilde{\sigma})d\tau = 1$, we have have

$$\int_0^\infty \tau^{-(\nu+N+1)}\exp(-\frac{\nu\sigma^2 + z^\top z}{2\tau^2})d\tau = \frac{1}{2}\Gamma(\frac{\nu+N}{2})(\frac{\nu\sigma^2 + z^\top z}{2})^{-(\nu+N)/2}.$$

Therefore, the probability density function of $z = y - \mu \in \mathbb{R}^N$ for $\mu = BAx \in \mathbb{R}^N$ is [1]

$$\int_0^\infty (2\pi\tau^2)^{-N/2}\exp(-\frac{z^\top z}{2\tau^2})f_{IG}(\tau|\nu, \sigma)d\tau$$

$$= \int_0^\infty (2\pi\tau^2)^{-N/2}\exp(-\frac{z^\top z}{2\tau^2}) \cdot \frac{2}{\Gamma(\frac{\nu}{2})}(\nu\sigma^2/2)^{\frac{\nu}{2}}\tau^{-\nu-1}\exp(-\frac{\nu\sigma^2}{2\tau^2})d\tau$$

$$= (2\pi)^{-N/2}\frac{2}{\Gamma(\nu/2)}(\frac{\nu\sigma^2}{2})^{\nu/2}\int_0^\infty \tau^{-(\nu+N+1)}\exp(-\frac{\nu\sigma^2 + z^\top z}{2\tau^2})d\tau$$

$$= (2\pi)^{-N/2}\frac{2}{\Gamma(\nu/2)}(\frac{\nu\sigma^2}{2})^{\nu/2} \cdot \frac{1}{2}\Gamma(\frac{\nu+N}{2})(\frac{\nu\sigma^2 + z^\top z}{2})^{-(\nu+N)/2} \qquad (9.18)$$

$$= \frac{\Gamma(\frac{\nu+N}{2})}{\Gamma(\frac{\nu}{2})}\frac{1}{(\nu\pi)^{N/2}\sigma^N}(1 + \frac{z^\top z}{\nu\sigma^2})^{-(\nu+N)/2}. \qquad (9.19)$$

Equation (9.19) is called the multivariate student-t distribution with parameters (ν, σ).

We have created the Stan code `model16.stan` to set the prior distributions obtained so far, calculate the posterior distribution, and obtain the WBIC values. The code utilizes the `functions` block to define custom functions, specifically for the definition of Leung-Drton's prior distribution.

model16.stan

```
1  functions {
2    real ld_diag_lpdf(real x, int i, int k, real c0) {
3      return (k-i)*log(x) - square(x)/(2*c0);
4    }
5  }
6  data {
7    int n;
8    int M;
9    int N;
```

```
10    int H;
11    matrix[n, M] X;
12    matrix[n, N] Y;
13    real<lower=0> beta;
14  }
15  transformed data {
16    int ntrap = M*H;
17  }
18  parameters {
19    real<lower=0> sigma;
20    real<lower=0> nu;
21    vector<lower=0>[H] diags;
22    vector[ntrap] lowtrap;
23    matrix[H, N] BhatT;
24    vector<lower=0>[M] lambda;
25  }
26  transformed parameters {
27    matrix[n, N] mu;
28    matrix[M, H] L;
29    {
30    int idx;
31    idx=0;
32    L = rep_matrix(0, M, H);
33    for (col in 1:H) {
34      L[col, col] = diags[col];
35      for (r in (col+1):M) {
36        idx+=1;
37        L[r, col] = lowtrap[idx];
38      }
39    }
40    mu = diag_post_multiply(X, lambda)*L*BhatT;
41    }
42  }
43  model {
44    lowtrap ~ normal(0,1);
45    for (i in 1:H)
46      diags[i] ~ ld_diag(i, H, 1);
47    to_vector(BhatT) ~ normal(0,1);
48    lambda ~ normal(0, 2);
49    nu ~ gamma(2, 0.1);
50    sigma ~ cauchy(0,1);
51    for(j in 1:n){
52      for (i in 1:N)
53        target += beta*student_t_lpdf(Y[j,i]|nu, mu[j,i], sigma);
54    }
55  }
56  generated quantities{   // This block is for calculating the value of
57      WBIC.
58    matrix[n,N] log_lik;
59    for (j in 1:n) {
60      for (i in 1:N)
61        log_lik[j,i] = student_t_lpdf(Y[j,i]| nu, mu[j,i], sigma);
62    }
63  }
```

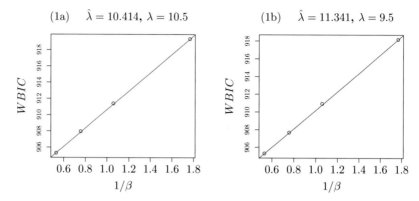

(1a) $\hat{\lambda} = 10.414$, $\lambda = 10.5$ (1b) $\hat{\lambda} = 11.341$, $\lambda = 9.5$

Fig. 9.3 For the three cases in Example 79, we calculated the WBIC values for multiple β values and plotted the relationship between the corresponding WBIC values using a straight line. The plots represent cases 1a, 1b, and 3, respectively

Example 79 Using the `model16.stan` code, we executed the following R code for cases 1a, 1b, and 3. We calculated the WBIC values for multiple β values and plotted the relationship between the corresponding WBIC values using a straight line (Fig. 9.3). From the slope of the line, we obtained estimated values of λ as shown in the table below. Note that the `ben_files.stan` used for data generation is identical to "Listing 2: Stan program for simulating data from the reduced-rank regression model" in the literature [3].[3]

CASE	M	N	H	r	λ's Estimate $\hat{\lambda}$	λ's Theoretical Value
1a	5	5	4	2	10.414	10.5
1b	5	5	3	2	11.341	9.5
3	3	4	3	3	6.329	6.0

∎

```
1     wbic <- function(log_likelihood) - mean(rowSums(log_likelihood))
2     library(rstan)
3     ## Data Generation Process
4     n <- 200
5     M <- 4
6     N <- 3
7     r <- 2
8     data_list <- list(n=100, p=M, c=N, k=r)
9     fit <- stan(file = "ben_files.stan", data = data_list, chain=1,
10    algorithm='Fixed_param')
11    fit_ss <- extract(fit, permuted = TRUE)
12    X <- fit_ss$X[1,,]
13    Y <- fit_ss$Y[1,,]
14    ## Calculate the WBIC values for multiple betas
```

---

[3] It is also available from the website of this series: https://bayesnet.org/books_jp.

```
15    WBIC <- NULL
16    B.seq <- c(3, 5, 7, 10)
17    for(b in B.seq){
18        data_list <- list(n=n,M=3,H=3,N=4,X=X,Y=Y, beta=b/log(n))
19        fit <- stan(file="model16.stan",data=data_list)
20        wbic.value <- wbic(extract(fit)$log_lik)
21        WBIC <- c(WBIC, wbic.value)
22    }
23    beta.1<-B.seq/log(n)
24    u<-1/beta.1; uu<-u-mean(u)
25    v<-WBIC; vv<-v-mean(v)
26    slope<-sum(uu*vv)/sum(uu*uu)
27    intercept<-mean(v)-slope*mean(u)
28    plot(u,v, xlab="1/beta",ylab="WBIC")
29    abline(a=intercept, b=slope, col="red")
```

9.4 Application to Gaussian Mixture Models

Suppose there are multiple Gaussian distributions (we assume $H \geq 1$), and one of them is randomly chosen (with probabilities $\pi_1, \ldots, \pi_H \geq 0$, and $\sum_{h=1}^{H} \pi_h = 1$), and the N-dimensional data $x \in \mathbb{R}^N$ is generated according to this Gaussian distribution $N(\mu_h, \sigma_h^2)$. For simplicity of discussion, we will assume that $\sigma_h^2 = 1$:

$$p(x|\theta) = \sum_{h=1}^{H} \pi_h s_h(x \mid \mu_h)$$

with

$$\theta = \{(\pi_h, \mu_h) \mid h = 1, \ldots, H\},$$

where $s_h(x|\mu_h, \sigma_h^2)$ is the probability density function of the normal distribution with mean μ_h and variance σ_h^2. Also, we assume that the above statistical model is realizable ($q(x) = p(x \mid \theta_*)$), and we denote the true H as H_*, and the true θ as

$$\theta_* = \{(\pi_h^*, \mu_h^*) \mid h = 1, \ldots, H_*\}.$$

Regarding the equation

$$\log \frac{p(x \mid \theta_*)}{p(x \mid \theta)} = -\log \left(1 + \frac{p(x \mid \theta) - p(x \mid \theta_*)}{p(x \mid \theta_*)}\right), \qquad (9.20)$$

for arbitrary $\epsilon > 0$, there exist $\theta \neq \theta_*$ with $|\theta - \theta_*| < \epsilon$ and $x \in R^N$ such that

$$\left|\frac{p(x\mid\theta)-p(x\mid\theta_*)}{p(x\mid\theta_*)}\right|>1.$$

Thus, the convergence radius of $\log(\cdot)$ is 1, and we cannot write (9.20) in the form of (5.2).

For example, if $N=1$, $H_*=1$, $H=2$, and μ,

$$\frac{p(x\mid\theta)-p(x\mid\theta_*)}{p(x\mid\theta_*)}=\frac{\pi_1\exp\{-\frac{(x-\theta)^2}{2}\}+(1-\pi_1)\exp\{-\frac{x^2}{2}\}-\exp\{-\frac{(x-\theta_*)^2}{2}\}}{\exp\{-\frac{(x-\theta_*)^2}{2}\}}$$

$$=\pi\{\exp(\frac{x\mu_1}{2}-\frac{\mu_1^2}{2})-1\}$$

becomes, and (9.20) does not become, a holomorphic function with values in $L^2(q)$. This means that it does not satisfy Assumption 4 that we have been assuming. However, even in that case, it has been proven that there exists an always positive C^∞ (not holomorphic) $a(u)$ such that [12]

$$K(g(u))=a(u)u_1^{2k_1}\ldots u_d^{2k_d}.$$

From now on, we will first calculate the value of WBIC(H) for the observed values $x_1,\ldots,x_n\in\mathbb{R}^N$ and the candidate H, and find the H that minimizes it.

Example 80 We shall consider data $x_1,\ldots,x_n\in\mathbb{R}^n$ with dimension $N=2$. With $\mu=(1,1)$, and the identity matrix of size $N=2$ as Σ, we generate data $x_1,\ldots,x_n$ $(n=200)$ each from the two kinds of Gaussian distributions $N(-\mu,\Sigma)$ and $N(\mu,\Sigma)$, and calculate the value of the Gaussian mixture WBIC for $H=1,2,3,4$, considering $\pi_1^*=\pi_2^*=1/2$, $H_*=2$, and the H that minimizes that value is used as the estimate of H_*. Similarly, we generate data $x_1,\ldots,x_n$ $(n=300)$ each from the three kinds of Gaussian distributions $N(-2\mu,\Sigma)$, $N(0,\Sigma)$, and $N(2\mu,\Sigma)$, and calculate the value of the Gaussian mixture WBIC for $H=1,2,3,4$, considering $\pi_1^*=\pi_2^*=\pi_3^*=1/3$, $H_*=3$, and the H that minimizes that value is used as the estimate of H_*. Also, experiments were conducted for $\beta=1/\log n,10/\log n,100/\log n,250/\log n$. The results are shown in Table 9.2. In the case of $H_*=2$, it is correctly estimated. In the case of $H_*=3$, the value of WBIC is smaller for $H=4$. When H is large, unless the sample size n is large, the problem of estimating H_* is not easy. ∎

In addition, the learning coefficient, when $H\geq H_*$, is bounded from above by Proposition 43 as

$$\lambda\leq\frac{1}{2}(NH_*+H-1). \tag{9.21}$$

In fact, if we define

$$\pi_h=\begin{cases}\pi_h^*,\ 1\leq h\leq H_*\\0,\ \ otherwise\end{cases},\quad\mu_h=\begin{cases}\mu_h^*,\qquad\quad 1\leq h\leq H_*\\arbitrary,\ otherwise,\end{cases}$$

Table 9.2 Computing WBIC for $N = 1$, $H = 2, 3$, and $H = 1, 2, 3, 4$

H_*	2 ($n = 200$)				3 ($n = 600$)			
$\beta \backslash H$	1	2	3	4	1	2	3	4
$1/\log n$	753.44	662.36	664.05	665.28	3168.2	2239.4	2178.8	2178.9
$10/\log n$	748.80	650.64	650.77	650.98	3163.5	2227.4	2157.897	2157.887
$100/\log n$	748.33	649.46	648.97	649.01	3163.0	2226.2	2156.0	2155.6
$250/\log n$	748.30	649.38	648.83	648.85	3163.0	2226.1	2156.0	2155.5
Upper bound of λ	2	5/2	3	7/2	3	7/2	4	9/2

we have a dimension of $NH + H - 1$, but there are redundant parameters of $N(H - H_*)$, and by Proposition 43, the learning coefficient is bounded above by (9.21). The upper bound in the case of Example 80 is recorded in the lower row of Table 9.2. The learning coefficient has been studied not only for the upper bound but also for the lower bound. If they are close, it can be good information. Also, it can be numerically estimated using the method shown at the end of Sect. 8.1.

model17.stan

```
data {
  int<lower=1> K;          // number of mixture components
  int<lower=1> N;          // number of data points
  array N vector[2] y;              // observations
  real beta;
}
parameters {
  simplex[K] theta;        // mixing proportions
  vector[2] mu[K];   // locations of mixture components
}
transformed parameters{
  vector[K] log_theta = log(theta);  // cache log calculation
}
model {
  mu ~ multi_normal(rep_vector(0.0,2), diag_matrix(rep_vector(1.0,2)));
  for (n in 1:N) {
    vector[K] lps = log_theta;
    for (k in 1:K)
      lps[k] += multi_normal_lpdf(y[n] | mu[k], diag_matrix(rep_vector
  (1.0,2)));
    target += beta*log_sum_exp(lps);
  }
}
generated quantities{
  vector[N] log_lik;
  for (n in 1:N) {
    vector[K] lps = log_theta;
    for (k in 1:K)
```

```
29      lps[k] += multi_normal_lpdf(y[n] | mu[k], diag_matrix(rep_vector
30      (1.0,2)));
31      log_lik[n] = log_sum_exp(lps);
32    }
33  }
```

```
1   wbic <- function(log_likelihood) - mean(rowSums(log_likelihood))
2   library(rstan)
3   b.seq <- c(1,10,100,250)
4   K.seq <- c(1,2,3,4)
5
6   x <- list()
7   #n <- 100     # Data generation K=2
8   #for(i in 1:n) x[[i]]=c(rnorm(1,-1,1),rnorm(1,-1,1))
9   #for(i in (n+1):(2*n)) x[[i]]=c(rnorm(1,1,1),rnorm(1,1,1))
10  n <- 200     # Data generation K=3
11  for(i in 1:n) x[[i]] <- c(rnorm(1,-2,1),rnorm(1,-2,1))
12  for(i in (n+1):(2*n)) x[[i]] <- c(rnorm(1,2,1),rnorm(1,2,1))
13  for(i in (2*n+1):(3*n)) x[[i]] <- c(rnorm(1,0,1),rnorm(1,0,1))
14
15  WBIC <- NULL
16  for(b in b.seq)for(k in K.seq){
17    data_list <- list(K = k, N = length(x), y=x, beta=b/log(n))
18    fit <- stan(file = "model17.stan", data = data_list, warmup = 2500,
19        seed = 1, iter=5000)
20    mm <- rstan::extract(fit)
21    wbic.1 <- wbic(mm$log_lik)
22    WBIC <- c(WBIC,wbic.1)
23  }
24  WBIC
```

9.5   Non-informative Prior Distribution

We assume that the determinant $\det I(\theta)$ of the Fisher information matrix $I(\theta)$ at each $\theta \in \Theta$ is non-negative definite and the integral $\int_{\Theta} \sqrt{\det I(\theta)} d\theta$ is defined. In obtaining $I(\theta)$, we took the expectation value with $q(x)$ in Chap. 5, but here we take the expectation value with $p(x|\theta)$. Furthermore, when this integral has a finite positive value, we call

$$\varphi_*(\theta) := \frac{\sqrt{\det I(\theta)}}{\int_{\Theta} \sqrt{\det I(\theta')} d\theta'}$$

Jeffreys prior distribution. When $K(\theta) = 0$, $\varphi(\theta) = 0$, so the discussion in Sect. 9.2 does not hold. By applying Jeffreys prior distribution $\varphi = \varphi_*$, (6.23) can be written as

$$\mathbb{E}_{X_1 \dots X_n}[F_n] = n \mathbb{E}_X[-\log p(X|\theta_*)] + \frac{d}{2}\log\frac{n}{2\pi e} + \log\int_\Theta \sqrt{\det I(\theta)}d\theta + o(1).$$

$$(9.22)$$

Example 81 For the log-likelihood of a normal distribution with variance 1 and known mean μ,

$$f(x|\mu) = \frac{1}{\sqrt{2\pi}}\exp\{-\frac{(x-\mu)^2}{2}\},$$

$L := \sum_{i=1}^n \log f(x_i|\mu)$, and taking the derivative with respect to μ gives $\sum_{i=1}^n (x_i - \mu)$, and its square mean is n, so $I = 1$. Therefore, if Θ is not bounded, such as $\Theta = [-2, 2]$, Jeffreys prior distribution φ_* is not determined. ∎

Example 82 Consider a random variable X that takes on the value 1 with probability θ and the value 0 with probability $1 - \theta$. Let $x_1, \dots, x_n \in \mathcal{X} = \{0, 1\}$ be independent realizations of X. For $L := \log\{\theta^k(1-\theta)^{n-k}\}$, we have $L' = \frac{k}{\theta} - \frac{n-k}{1-\theta} = \frac{k-n\theta}{\theta(1-\theta)}$.

The square mean of this is $\mathbb{E}[\left(\frac{k-n\theta}{\theta(1-\theta)}\right)^2] = \frac{n}{\theta(1-\theta)}$, so $I(\theta) = \frac{1}{\theta(1-\theta)} > 0$. Therefore, Jeffreys prior distribution is proportional to $\varphi_*(\theta) \propto \theta^{-1/2}(1-\theta)^{-1/2}$. This corresponds to setting $a = b = 1/2$ in Eq. (2.19). ∎

In Bayesian statistics, a prior distribution that is used when there is no particular information available for determining the prior distribution is sometimes called a *non-informative prior*. In Eq. (2.19), for example, $a = b = 1$ (a uniform distribution) is often used in practice as a non-informative prior. However, a uniform distribution over $0 \le \theta \le 1$ is not a uniform distribution when the variable is transformed to $t := \sqrt{\theta}$. On the other hand, Jeffreys prior does not suffer from this problem, satisfying the so-called *invariance property*. In some cases, a prior distribution satisfying invariance is defined as a non-informative prior.

Example 83 In the case of Jeffreys prior, let's demonstrate that when a random variable X takes the value 1 with probability θ and the value 0 with probability $1 - \theta$, if we let $\theta = t^2$ and consider X taking the value 1 with probability t^2 and the value 0 with probability $1 - t^2$, Jeffreys prior distribution remains the same distribution. We shall denote Jeffreys prior distribution for each as φ_*^Θ and φ_*^T, respectively. Let $x_1, \dots, x_n \in \mathcal{X} = \{0, 1\}$ be independent realizations of the random variable X. For $L := \log\{t^{2k}(1-t^2)^{n-k}\}$, we have

$$L' = \frac{2k}{t} - \frac{2t(n-k)}{1-t^2} = \frac{2(k-nt^2)}{t(1-t^2)}, \quad \mathbb{E}[(L')^2] = \frac{4n}{1-t^2}, \quad I(t) = \frac{4}{1-t^2}.$$

Hence, Jeffreys prior distribution is $\varphi_*^T(t) \propto (1-t^2)^{-1/2}$. Here, since $\theta = t^2$, we have $d\theta = 2tdt$, so

$$\frac{d\theta}{\sqrt{\theta(1-\theta)}} = \frac{2tdt}{\sqrt{t^2(1-t^2)}} = \frac{2dt}{\sqrt{1-t^2}}.$$

This implies that, whether we take the parameter as θ or t, for the integrable subset $T' \subseteq T$ with $t \in [0, 1]$ and $\Theta' = \{t^2 | t \in T'\} \subseteq \Theta$, we have

$$\int_{T'} \varphi_*^T(t)dt = \int_{\Theta'} \varphi_*^{\Theta}d\theta.$$

∎

Proposition 47 *When applying Jeffreys prior distribution, one of the following occurs:*

1. $\lambda = d/2$ *and* $m = 1$
2. $\lambda > d/2$.

Proof Refer to the appendix at the end of the chapter.

Appendix: Proof of Proposition

Proof of Proposition 40

For $p = 0, 1$, we have

$$W_n^{(p)} := \int_{B(\epsilon_n, \theta_*)} \{-\sum_{i=1}^{n} \log \frac{p(x_i|\theta)}{p(x_i|\theta_*)}\}^p \{\prod_{i=1}^{n} \frac{p(x_i|\theta)}{p(x_i|\theta_*)}\}^{\beta} \varphi(\theta)d\theta$$

$$= \sum_{\alpha} \int_{[0,1]^d} du \{nu^{2k} - \sqrt{n}u^k \xi_n(u)\}^p \exp\{-n\beta u^{2k} + \sqrt{n}\beta u^k \xi_n(u)\}|u^h|b_{\alpha}(u).$$

We evaluate the value of

$$\mathcal{E}_{\beta}[\sum_{i=1}^{n} \log \frac{p(x_i|\theta_*)}{p(x_i|g(u))}] = \frac{W_n^{(1)} + o_P(\exp(-\sqrt{n}))}{W_n^{(0)} + o_P(\exp(-\sqrt{n}))}, \qquad (9.23)$$

where we have performed a variable transformation with $t = n\beta u^{2k}$. Using $\exp(\sqrt{\beta t}\xi_n(u)) = 1 + \sqrt{\beta t}\xi_n(u) + O_P(\beta)$ as $\beta \to 0$, and excluding the term $o_P(\frac{(\log n\beta)^{m-1}}{(n\beta)^{\lambda}})$, we obtain

$$W_n^{(p)}$$

$$\sim \frac{(\log n\beta)^{m-1}}{(n\beta)^\lambda} \sum_{\alpha \in A} \int_{[0,1]^d} \int_0^\infty dt\, t^{\lambda-1} e^{-t} \left\{ \frac{t - \sqrt{\beta t}\xi_n(u)}{\beta} \right\}^p \{1 + \sqrt{\beta t}\xi_n(u)\} du_\alpha^*$$

$$= \begin{cases} \dfrac{(\log n\beta)^{m-1}}{(n\beta)^\lambda}\left[\Gamma(\lambda)C^{(0)} + \sqrt{\beta}\Gamma(\lambda + \dfrac{1}{2})C^{(1)} \right], & p = 0 \\[2mm] \dfrac{(\log n\beta)^{m-1}}{(n\beta)^\lambda \beta}\left[\Gamma(\lambda+1)C^{(0)} + \sqrt{\beta}\{\Gamma(\lambda + \dfrac{3}{2}) - \Gamma(\lambda + \dfrac{1}{2})\}C^{(1)} \right], & p = 1 \end{cases}$$

where we define

$$C^{(p)} := \left(\sum_{\alpha \in A} \int_{[0,1]^d} \xi_n(u)^p du_\alpha^* \right), \quad p = 0, 1, \quad \Gamma(\lambda) := \int_0^\infty t^{\lambda-1} \exp(-t) dt$$

We shall define $Y := C^{(1)}/C^{(0)}$. Then, for constants a, b, c, d with $a \neq 0$, as $\beta \to 0$, we have

$$\frac{c + \sqrt{\beta}d}{a + \sqrt{\beta}b} = \frac{c}{a} + \sqrt{\beta}\left(\frac{ad - bc}{a^2} \right) + O(\beta).$$

Setting $a = \Gamma(\lambda)$, $b = \Gamma(\lambda + 1/2)Y$, $c = \Gamma(\lambda + 1)$, and $d = (\lambda - 1/2)\Gamma(\lambda + 1/2)Y$, we obtain

$$\frac{c}{a} = \frac{\Gamma(\lambda + 1)}{\Gamma(\lambda)} = \lambda$$

and

$$ad - bc = \Gamma(\lambda) \cdot (\lambda - \frac{1}{2})\Gamma(\lambda + \frac{1}{2})Y - \Gamma(\lambda + \frac{1}{2})Y \cdot \Gamma(\lambda + 1) = -\frac{1}{2}\Gamma(\lambda)\Gamma(\lambda + \frac{1}{2})Y.$$

Therefore, Eq. (9.23) can be written as

$$\mathcal{E}_\beta[\sum_{i=1}^n \log \frac{p(x_i|\theta_*)}{p(x_i|g(u))}] = \frac{\lambda}{\beta} + U_n\sqrt{\frac{\lambda}{2\beta}} + O_P(1),$$

where we define

$$U_n := -\frac{Y\Gamma(\lambda + 1/2)}{\sqrt{2\lambda}\Gamma(\lambda)}.$$

Substituting $\beta = \beta_0/\log n$ into this equation, we obtain Eq. (9.3). From the definition of $\xi_n(u)$, we get $\mathbb{E}_{X_1 \ldots X_n}[Y] = 0$ and $\mathbb{E}_{X_1 \ldots X_n}[U_n] = 0$. Furthermore, since the sequence $\xi_1, \xi_2, \ldots$ converges in distribution to the Gaussian process ξ, the sequence of random variables $U_1, U_2, \ldots$ also converges in distribution to a normal distribution. This completes the proof. ∎

Proof of Proposition 46

In order to prove this proposition, we use the following lemma.

Lemma 1 *Let U be a neighborhood of $\theta_* \in \Theta_*$, and $l, m, n \geq 1$. Suppose T_1, T_2, T are mappings from $U \to \mathbb{R}^{l \times m}$, $U \to \mathbb{R}^{l \times n}$, $U \to \mathbb{R}^{m \times n}$ matrices respectively, and $\| \cdot \|$ are their norms. Then,*

1. *If $P \in \mathbb{R}^{m \times m}$, $Q \in \mathbb{R}^{n \times n}$ are invertible, then when considering PTQ as a function assigning the matrix $PT(u)Q \in \mathbb{R}^{m \times n}$ to elements u of U, there exist $\alpha, \beta > 0$ such that $\alpha \|T\| \leq \|PTQ\| \leq \beta \|T\|$.*
2. *If T is bounded ($\|T\| < \infty$), then there exist $\alpha, \beta > 0$ such that $\alpha(\|T_1\|^2 + \|T_2\|^2) \leq \|T_1\|^2 + \|T_2 + T_1T\|^2 \leq \beta(\|T_1\|^2 + \|T_2\|^2)$.*

Proof of Lemma 1: Refer to the next section.

Proof of Proposition 46[4]: Firstly, let $\theta = (A, B)$, $\theta_* = (A_*, B_*)$, and $S := BA - B_*A_*$. Then

$$K(\theta) = \frac{1}{2}\mathbb{E}_{XY}\left[\|Y - BAX\|^2 - \|Y - B_*A_*X\|^2 \right]$$

$$= \frac{1}{2}\mathbb{E}_{XY}\left[\|Y - B_*A_*X + (B_*A_* - BA)X\|^2 - \|Y - B_*A_*X\|^2 \right]$$

$$= \frac{1}{2}\int_{\mathcal{X}} \|Sx\|^2 q(x)dx$$

(considering $Y - B_*A_*X \sim N(0, 1)$), the problem of finding the poles of $\int_\Theta K(\theta)^z \varphi(\theta)d\theta$ is equivalent to finding the poles of $\int_\Theta \|S\|^{2z}\varphi(\theta)d\theta$. To show this, let $S = (s_{jk})$. Then, there exists a matrix X such that

$$K(\theta) = \frac{1}{2}\sum_i \int_{\mathcal{X}} (\sum_j s_{ij}x_j)^2 q(x)dx = \frac{1}{2}\sum_i \sum_j \sum_k s_{ij}s_{ik}\int_{\mathcal{X}} x_j x_k q(x)dx = \frac{1}{2}Tr(SXS^T)$$

and without loss of generality, X can be assumed to be positive definite. Then, there exist constants $c_1, c_2 > 0$ such that

$$c_1\|S\|^2 = c_1 Tr(SS^T) \leq K(\theta) \leq c_2 Tr(SS^T) = c_2\|S\|^2.$$

(For example, take the minimum and maximum eigenvalues of the matrix X as $2c_1, 2c_2$, respectively.) Therefore, the equivalence has been shown.

---

[4] An attempt has been made to derive this in a simple way so that non-experts can understand.

Let $A = (a_{i,j}) \in \mathbb{R}^{H \times M}$, $B = (b_{i,j}) \in \mathbb{R}^{N \times H}$, and assume that the rank of $B_* A_*$ is r, and let $\| \cdot \|$ denote the Frobenius norm of the matrix (the square root of the sum of squares of the components). First, since the rank of $B_* A_*$ is r, there exist regular matrices $P \in \mathbb{R}^{N \times N}$, $Q \in \mathbb{R}^{M \times M}$ such that

$$P^{-1} B_* A_* Q^{-1} = \begin{bmatrix} I_r & 0 \\ 0 & 0 \end{bmatrix}$$

and we have

$$B' := P^{-1} B = \begin{bmatrix} B_1 & B_3 \\ B_2 & B_4 \end{bmatrix} \text{ and } A' := A Q^{-1} = \begin{bmatrix} A_1 & A_3 \\ A_2 & A_4 \end{bmatrix}.$$

Let $A_1 \in \mathbb{R}^{r \times r}$, $A_2 \in \mathbb{R}^{(H-r) \times r}$, $A_3 \in \mathbb{R}^{r \times (M-r)}$, $A_4 \in \mathbb{R}^{(H-r) \times (M-r)}$, $B_1 \in \mathbb{R}^{r \times r}$, $B_2 \in \mathbb{R}^{(N-r) \times r}$, $B_3 \in \mathbb{R}^{r \times (H-r)}$, and $B_4 \in \mathbb{R}^{(N-r) \times (H-r)}$. We can further transform

$$
\begin{aligned}
T := B' A' - \begin{bmatrix} I_r & 0 \\ 0 & 0 \end{bmatrix} &= \begin{bmatrix} B_1 & B_3 \\ B_2 & B_4 \end{bmatrix} \begin{bmatrix} A_1 & A_3 \\ A_2 & A_4 \end{bmatrix} - \begin{bmatrix} I_r & 0 \\ 0 & 0 \end{bmatrix} \\
&= \begin{bmatrix} C_1 & (C_1 + I_r - B_3 A_2) A_1^{-1} A_3 + B_3 A_4 \\ C_2 & (C_2 - B_4 A_2) A_1^{-1} A_3 + B_4 A_4 \end{bmatrix} \\
&= \begin{bmatrix} C_1 & C_1 A_1^{-1} A_3 + A_1^{-1} A_3 + B_3 A_4' \\ C_2 & C_2 A_1^{-1} A_3 + B_4 A_4' \end{bmatrix} = \begin{bmatrix} C_1 & C_1(A_3' - B_3 A_4') + A_3' \\ C_2 & C_2(A_3' - B_3 A_4') + B_4 A_4' \end{bmatrix}.
\end{aligned}
$$

Let $C_1 := B_1 A_1 + B_3 A_2 - I_r$, $C_2 := B_2 A_1 + B_4 A_2$, $A_4' := -A_2 A_1^{-1} A_3 + A_4$, and $A_3' := A_1^{-1} A_3 + B_3 A_4'$. Also, for any neighborhood $U_{(A', B')}$ of (A', B') such that

$$B' A' = \begin{bmatrix} I_r & 0 \\ 0 & 0 \end{bmatrix}.$$

According to Lemma 1 1, there exist $\alpha, \beta > 0$ such that $\alpha \| T \|^2 \le \| P T Q \|^2 \le \beta \| T \|^2$. Also, since

$$P T Q = P \left(B' A' - \begin{bmatrix} I_r & 0 \\ 0 & 0 \end{bmatrix} \right) Q = B A - B_* A_*,$$

let's define $\Psi := \| B A - B_* A_* \|^2$. Then, the poles of the integral

$$\int_{U_{(A', B')}} \Psi^z \varphi(\theta) d\theta$$

coincide with the poles of the integral

$$\int_{U_{(A', B')}} \left\| \begin{bmatrix} C_1 & C_1(A_3' - B_3 A_4') + A_3' \\ C_2 & C_2(A_3' - B_3 A_4') + B_4 A_4' \end{bmatrix} \right\|^{2z} \varphi d\theta$$

Furthermore, using Lemma 1 2, if we define

$$T_1 := \begin{bmatrix} C_1 \\ C_2 \end{bmatrix}, \quad T_2 := \begin{bmatrix} A_3' \\ B_4 A_4' \end{bmatrix}, \quad T := A_3' - B_3 A_4',$$

then the poles of the integral

$$\int_{U_{(A',B')}} \left\| \begin{bmatrix} C_1 & A_3' \\ C_2 & B_4 A_4' \end{bmatrix} \right\|^{2z} \varphi(\theta) d\theta$$

also coincide. Moreover, from $C_1 \in \mathbb{R}^{r \times r}$, $C_2 \in \mathbb{R}^{(N-r) \times r}$, and $A_3' \in \mathbb{R}^{r \times (M-r)}$, we can construct the matrix $C^{(0)} := [C_1, C_2^\top, A_3'] \in \mathbb{R}^{r \times (N+M-r)}$, $B^{(0)} := B_4 \in \mathbb{R}^{(N-r) \times (H-r)}$, and $A^{(0)} := A_4' \in \mathbb{R}^{(H-r) \times (M-r)}$. If we let $\Phi^{(0)} := \|C^{(0)}\|^2 + \|B^{(0)} A^{(0)}\|^2$, then the poles of the integral

$$\int_{U_{(A',B')}} \{\Phi^{(0)}(\theta)\}^z \varphi(\theta) d\theta$$

also coincide.

The following demonstrates that for each $s = 0, 1, \ldots, \min\{H - r, M - r\}$, there exist $C^{(s)} \in \mathbb{R}^{r \times (N+M-r)}$, $A^{(s)} \in \mathbb{R}^{(H-r-s) \times (M-r-s)}$, $B^{(s)} \in \mathbb{R}^{(N-r) \times (H-r-s)}$, $D_1^{(s)}, \ldots,$ $D_s^{(s)} \in \mathbb{R}^{1 \times (M-r-s)}$, $b_1^{(s)}, \ldots, b_s^{(s)} \in \mathbb{R}^{N-r}$ such that

$$\Phi^{(s)} := u_1^2 \ldots u_s^2 \left(\|C^{(s)}\|^2 + \sum_{i=1}^{s} \|b_i^{(s)}\|^2 + \| \sum_{i=1}^{s} b_i^{(s)} D_i^{(s)} + B^{(s)} A^{(s)} \|^2 \right) \quad (9.24)$$

To demonstrate that a similar $\Phi^{(s+1)}$ can be obtained by blowing up one component of $A^{(s)} \in \mathbb{R}^{(H-r-s) \times (M-r-s)}$, we without loss of generality consider the top-left component u_{s+1} of $A^{(s)}$ to be the one to be blown up. The other components of $A^{(s)}$, as well as the components of $C^{(s)}$ and $b^{(1)}, \ldots, b^{(s)}$, are transformed by multiplying by u_{s+1}. By factoring out u_{s+1} from $\Phi^{(s)}$, we can express $C^{(s)}$ as $C^{(s+1)}$, $b_i^{(s)}$ as $b_i^{(s+1)}$, and $\sum_{i=1}^{s} b_i^{(s)} D_i^{(s)} + B^{(s)} A^{(s)}$ in terms of suitable $a \in \mathbb{R}^{(H-r-s-1) \times 1}$, $\tilde{a} \in \mathbb{R}^{1 \times (M-r-s-1)}$, and $\tilde{A} \in \mathbb{R}^{(H-r-s-1) \times (M-r-s-1)}$ as follows:

$$\sum_{i=1}^{s} b_i^{(s+1)} [D_{i,1}, D_{i,2}, \ldots, D_{i,M-r-s}] + [b \, B^{(s+1)}] \begin{bmatrix} 1 & \tilde{a} \\ a & \tilde{A} \end{bmatrix}.$$

where we assume $B^{(s)} = [b, B^{(s+1)}]$ and $D_i^{(s)} = [D_{i,1}, D_{i,2}, \ldots, D_{i,M-r-s}]$. By defining the leftmost column as

$$b_{s+1}^{(s+1)} := b + B^{(s+1)} a + \sum_{i=1}^{s} b_i^{(s+1)} D_{i,1},$$

we can rewrite the remaining columns as

$$\sum_{i=1}^{s} b_i^{(s+1)} [D_{i,2}, \ldots, D_{i,M-r-s}] + b\tilde{a} + B^{(s+1)}\tilde{A}$$

$$= \sum_{i=1}^{s} b_i^{(s+1)} [D_{i,2}, \ldots, D_{i,M-r-s}] + \{b_{s+1}^{(s+1)} - B^{(s+1)}a - \sum_{i=1}^{s} b_i^{(s+1)} D_{i,1}\}\tilde{a} + B^{(s+1)}\tilde{A}$$

$$= \sum_{i=1}^{s} b_i^{(s+1)} \{[D_{i,2}, \ldots, D_{i,M-r-s}] - D_{i,1}\tilde{a}\} + b_{s+1}^{(s+1)}\tilde{a} + B^{(s+1)}\{\tilde{A} - a\tilde{a}\}$$

$$= \sum_{i=1}^{s+1} b_i^{(s+1)} D_i^{(s+1)} + B^{(s+1)} A^{(s+1)},$$

where we have set $A^{(s+1)} := \tilde{A} - a\tilde{a}$ and defined $D_i^{(s+1)} \in \mathbb{R}^{1\times(M-r-s-1)}$ as

$$D_i^{(s+1)} := [D_{i,2}, \ldots, D_{i,M-r-s}] - D_{i,1}\tilde{a}$$

for $1 \le i \le s$, and $D_{s+1}^{(s+1)} := \tilde{a}$. Therefore, we can express $\Phi^{(s+1)}$ in the form of Eq. (9.24) with s replaced by $s + 1$.

When $s = \min\{H - r, M - r\}$ is reached, the term $B^{(s)} A^{(s)}$ vanishes. However, this operation does not find all local coordinates of the branching manifolds. Choose some component of $C^{(s)}$ or $b^{(s)}$ as ν, and multiply all the other components of $C^{(s)}$, $b^{(s)}$, and $A^{(s)}$ by ν. This transformation forms a normal cross in this neighborhood, thus obtaining the desired local coordinates.

At this time, all components of $C^{(s)}$, $A^{(s)}$, and $b_1^{(s)}, \ldots, b_s^{(s)}$ in $\Phi^{(s)}$ are divided by ν, and $u_1 \ldots u_s$ becomes $u_1 \ldots u_s\nu$. Furthermore, at each s, $C^{(s)}$, $b^{(s)}$, and $A^{(s)}$ each have $(N + M - r)r, (N - r)s,$ and $(M - r - s)(H - r - s)$ elements, respectively, for a total of

$$l(s) := (N + M - r)r + s(N - r) + (M - r - s)(H - r - s)$$
$$= s^2 - (M + H - N - r)s + (N - H)r + MH$$

elements. The Jacobian when each variable is divided by the variable u_{s+1} is $u_{s+1}^{l(s)-1}$. The final Jacobian is $u_1^{l(0)-1} \ldots u_s^{l(s-1)-1}\nu^{l(s)-1}$, and $K(\theta) = u_1^2 \ldots u_s^2\nu^2$. Therefore, the poles of $u_1, \ldots, u_s, \nu$ are

$$-\frac{l(0)}{2}, \ldots, -\frac{l(s)}{2}.$$

Also, the number of times the former blow-up is performed, $s = 0, 1, \ldots, \min\{H - r, M - r\}$, is determined by when $C^{(s)}$ or $b^{(s)}$ is blown up, and can be freely chosen.

Finally, since $l(s)$ is a quadratic function for $0 \le s \le \min\{H - r, M - r\}$, if the minimum point is included in the interval, if $M + H - N - r$ is even (equivalent to $M + H + N + r$ being even), it is minimum at $s = (M + H - N - r)/2$, and if it is odd, it is minimum at $s = (M + H - N - r \pm 1)/2$. When $M + H < N - r$, it is minimum at $s = 0$, when $(M + H - N - r)/2 > H - r$ and $H < M$, it is minimum at $s = H - r$, and when $(M + H - N - r)/2 > M - r$ and $M < H$, it is minimum at $s = M - r$. And since $N \geq r$, $M + r > N + H$ means $M > H$ and $H + r > M + N$ means $H > M$, so $M + r > N + H$ and $H + r > M + N$ do not occur simultaneously.

Proof of Lemma 1

In general, for

$$
A = \begin{bmatrix} a_1 \\ \vdots \\ a_m \end{bmatrix} \in \mathbb{R}^{m \times l} , \quad B = [b_1, \ldots, b_n] \in \mathbb{R}^{l \times n}
$$

we have $\|AB\|^2 = \sum_i \sum_j \langle a_i, b_j \rangle^2 \leq \sum_i \sum_j \|a_i\|^2 \|b_j\|^2 = \|A\|^2 \|B\|^2$, where $\|\cdot\|$ is the Frobenius norm. Thus, there exist $\beta, \gamma > 0$ such that

$$
\|PTQ\|^2 \leq \|P\|^2 \|T\|^2 \|Q\|^2 \leq \beta \|T\|^2
$$

and

$$
\|T\|^2 = \|P^{-1} PT Q Q^{-1}\|^2 \leq \|P^{-1}\|^2 \|PTQ\|^2 \|Q^{-1}\|^2 \leq \gamma \|PTQ\|^2.
$$

If we set $\alpha := \gamma^{-1}$, the first item is satisfied. Next, if $\|T\|^2 < \infty$, then similarly, there exist $\beta, \gamma > 0$ that satisfy the following:

$$
\|T_1\|^2 + \|T_2 + T_1 T\|^2 \leq \|T_1\|^2 + 2\|T_2\|^2 + 2\|T_1 T\|^2 \leq \beta(\|T_1\|^2 + \|T_2\|^2),
$$

$$
\|T_2\|^2 \leq 2\{\|T_2 + T_1 T\|^2 + \| - T_1 T\|^2\} \leq 2\{\|T_2 + T_1 T\|^2 + \gamma \|T_1\|^2\}
$$

and

$$
\|T_1\|^2 + \|T_2\|^2 \leq 2\|T_2 + T_1 T\|^2 + (2\gamma + 1)\|T_1\|^2 \leq (2\gamma + 1)\{\|T_2 + T_1 T\|^2 + \|T_1\|^2\}.
$$

If we set $\alpha := (2\gamma + 1)^{-1}$, we have the second item as well.

Proof of Proposition 47

We want to show that the maximum pole of the function

$$\zeta(z) = \int_{\Theta} K(\theta)^z \sqrt{\det I(\theta)} d\theta$$

is at $-\lambda = -d/2$ (multiplicity $m = 1$) or $-\lambda < -d/2$. In general, there are multiple local coordinates. Choose arbitrary local coordinates $(\theta_1, \ldots, \theta_d)$ such that $K(\theta) = \theta_1^{2k_1} \theta_2^{2k_2} \ldots \theta_s^{2k_s}$, for some positive integers $k_1, k_2, \ldots, k_s$. Therefore, the log-likelihood ratio function can be written as

$$f(x, \theta) := \sum_{i=1}^{n} \log \frac{p(x_i|\theta_*)}{p(x_i|\theta)} = a(x, \theta) \theta_1^{k_1} \theta_2^{k_2} \ldots \theta_s^{k_s}.$$

Define

$$r_i(x, \theta) := \begin{cases} \frac{\partial a(x,\theta)}{\partial \theta_i} \theta_i + k_i a(x, \theta), & k_i \neq 0 \\ \frac{\partial a(x,\theta)}{\partial \theta_i}, & k_i = 0. \end{cases} \tag{9.25}$$

We have

$$\frac{\partial f(x, \theta)}{\partial \theta_i} = \begin{cases} r_i(x, \theta) \theta_1^{k_1} \ldots \theta_i^{k_i - 1} \ldots \theta_d^{k_d}, & k_i \neq 0 \\ r_i(x, \theta) \theta_1^{k_1} \ldots \theta_d^{k_d}, & k_i = 0 \end{cases}$$

Define the elements of the matrix $J(\theta) = (J_{i,j}(\theta))$ as

$$J_{i,j}(\theta) := \int_{\mathcal{X}} r_i(x, \theta) r_j(x, \theta) p(x|\theta) dx.$$

Then, we have

$$\sqrt{\det(I(\theta))} = \prod_{j:k_j \neq 0} \theta_j^{dk_j - 1} \sqrt{\det(J(\theta))}. \tag{9.26}$$

In fact, the elements of $I(\theta) = (I_{i,j}(\theta))$ are

$$I_{i,j}(\theta) = \int_{\mathcal{X}} \frac{\partial f(x, \theta)}{\partial \theta_i} \frac{\partial f(x, \theta)}{\partial \theta_j} p(x|\theta) dx$$

$$= \theta_1^{2k_1} \ldots \theta_s^{2k_d} \theta_i^{-I(k_i \neq 0)} \theta_j^{-I(k_j \neq 0)} \int_{\mathcal{X}} r_i(x, \theta) r_j(x, \theta) p(x|\theta) dx,$$

where $I(k_i \neq 0)$ takes the value 1 if $k_i \neq 0$ and 0 if $k_i = 0$. In this case, when the indices $(i, j) = (i_1, j_1), \ldots, (i_d, j_d)$ of $I_{i,j}(\theta)$ are chosen so that there are no duplicates in each of $i_1, \ldots, i_d$ and $j_1, \ldots, j_d$; their product can be written as

$$\prod_{h=1}^{d} I_{i_h, j_h}(\theta) = \prod_{l:k_l \neq 0} \theta_l^{2dk_l-2} \prod_{h=1}^{d} J_{i_h, j_h}(\theta). \tag{9.27}$$

Therefore, (9.26) is obtained. Moreover, when $k_1 \neq 0, k_2, \ldots, k_d = 0$, $K(\theta) = \theta_1^{2k_1}$, from (9.26), we have

$$\sqrt{\det I(\theta)} = \theta_1^{dk_1-1} \sqrt{\det J(\theta)} = c_1(\theta_1, \ldots, \theta_d) \theta_1^{dk_1-1},$$

where c_1 is an analytic function. Thus, we have

$$\zeta(z) = \int_\Theta K(\theta)^z \sqrt{\det I(\theta)} d\theta_1 \ldots d\theta_d$$
$$= \int_\Theta c_1(\theta_1, \ldots, \theta_d) \theta_1^{2k_1 z} \theta_1^{dk_1-1} d\theta_1 \ldots d\theta_d = \frac{h(z)}{k_1(2z+d)},$$

where $h(z)$ is an analytic function of z. Therefore, if a pole of $\zeta(z)$ exists, it is at $z = -d/2$ with multiplicity $m = 1$. In the general case where there are more than two i for which $k_i \neq 0$, we perform the following coordinate transformation. For simplicity of discussion, assume $k_1, \ldots, k_s > 0, k_{s+1}, \ldots, k_d = 0$, and $K(\theta) = \theta_1^{k_1} \ldots \theta_s^{k_s}$. Furthermore, with respect to $K(\theta)$, the property of normal crossing is preserved even when we change the local coordinates to

$$\theta_1 = u_1, \quad \theta_2 = u_1 u_2, \quad \ldots, \quad \theta_s = u_1 \ldots u_s$$

via a blow-up transformation. We denote this transformation as $g(u)$, where $u = (u_1, \ldots, u_s)$. If we define $\sigma_j := k_1 + \cdots + k_j$, then we can express

$$K(g(u)) = \prod_{j=1}^{s} u_j^{2\sigma_j}.$$

Also, since $\dfrac{\partial \theta_i}{\partial u_j}$ forms a lower triangular matrix, the Jacobian of the transformation is

$$|g'(u)| = \prod_{i=1}^{s} \frac{\partial \theta_i}{\partial u_i} = \prod_{j=1}^{s} u_j^{s-j}.$$

Furthermore, if for any $1 \leq i \leq s-1$, $u_i = 0$, then $\theta_h = 0$ for $h = s-1, s$. From (9.25), $r_h(x, \theta) = k_h a(x, \theta)$ holds for $h = s-1, s$. Thus, the jth column of the $h = s-1, s$ rows of $J(\theta)$ are respectively

$$k_h \int_x a(x, \theta) \{ \frac{\partial a}{\partial \theta_j} \theta_j + k_j a(x, \theta) \} p(x|\theta) dx,$$

which means that the $s - 1$ and s rows are proportional, causing the rank of the matrix $J(\theta)$ to decrease by 1. Therefore, $\det J(g(u)) = 0$. Also, since $J(g(u))$ is an analytic function, using an analytic function $c_2(u_1, \ldots, u_s)$, we can express

$$\det J(g(u)) = c_2(u_1, \ldots, u_s)u_1^2 \ldots u_{s-1}^2,$$

which means

$$\sqrt{\det I(g(u))} = c_2(u_1, \ldots, u_s) \prod_{l=1}^{s-1} u_l \cdot \prod_{j=1}^{s} \{\prod_{h=1}^{j} u_h\}^{dk_j - 1}$$

$$= c_2(u_1, \ldots, u_s)u_s^{d\sigma_s - 1} \prod_{j=1}^{s-1} u_j^{d\sigma_j - s + j}.$$

Therefore,

$$u^{2kz}\sqrt{\det I(g(u))}|g'(u)| = u_s^{2\sigma_s z + d\sigma_s - 1} \prod_{j=1}^{s-1} u_j^{2\sigma_j z + d\sigma_j}.$$

When integrating with respect to each of $u_1, \ldots, u_s$, the poles of $\zeta(z)$ at u_j ($1 \le j \le s - 1$), u_s are understood to be respectively

$$-\frac{d\sigma_j + 1}{2\sigma_j}, \quad -\frac{d}{2}.$$

∎

Exercises 87–100

87. Show that there exists an inverse temperature $\beta > 0$ such that the free energy F_n and $WBIC_n$ are equal.
88. Show that for $p = 0, 1$,

$$W_n^{(p)} := \sum_\alpha \int_{[0,1]^d} du_\alpha \left\{-nu^{2k} + \sqrt{n}u^k\xi_n(u)\right\}^p \exp\{-n\beta u^{2k} + \sqrt{n}\beta u^k\xi_n(u)\}|u^h|b_\alpha(u)$$

can be represented as

$$W_n^{(p)}$$

$$\sim \frac{(\log n\beta)^{m-1}}{(n\beta)^\lambda} \sum_{\alpha \in A} \int_{[0,1]^d} \int_0^\infty dt\, t^{\lambda-1} e^{-t} \left\{ \frac{t - \sqrt{\beta t} \xi_n(u)}{\beta} \right\}^p \{1 + \sqrt{\beta t} \xi_n(u)\} du_\alpha^*$$

$$= \begin{cases} \dfrac{(\log n\beta)^{m-1}}{(n\beta)^\lambda} \left[\Gamma(\lambda) C^{(0)} + \sqrt{\beta} \Gamma\left(\lambda + \frac{1}{2}\right) C^{(1)} \right], & p = 0 \\[2ex] \dfrac{(\log n\beta)^{m-1}}{(n\beta)^\lambda \beta} \left[\Gamma(\lambda + 1) C^{(0)} + \sqrt{\beta} \{\Gamma(\lambda + \frac{3}{2}) - \Gamma(\lambda + \frac{1}{2})\} C^{(1)} \right], & p = 1, \end{cases}$$

Here,

$$C^{(p)} := \left(\sum_{\alpha \in A} \int_{[0,1]^d} \xi_n(u)^p du_\alpha^* \right), \quad p = 0, 1, \quad \Gamma(\lambda) := \int_0^\infty t^{\lambda-1} \exp(-t) dt.$$

89. In the proof of Proposition 42, why does the lower bound of $Z(n)n^{d/2}$ converging to a positive value contradict with the first point of Proposition 41?
90. In the derivation of Proposition 44, where are each of the three conditions for regularity used?
91. In Example 77, derive Eqs. (9.12) and (9.15). Also, how is Eq. (9.16) derived from a Jacobian?
92. Derive each value in Table 9.1.
93. In the procedure for estimating λ in Example 78, what would the estimated value of λ be if we use the values from 10, ..., 60 divided by $\log n$ as β instead of 100, ..., 600? Also, which variable in the program represents the estimated value of λ?
94. When $M = N = 1$ and $H_* = 0$, prove the following inequality for when $\sqrt{H}$ is an integer. Also, when does the equality hold? Here, $[a]$ is the largest integer not exceeding $a > 0$:

$$\frac{[\sqrt{H}]^2 + [\sqrt{H}] + H}{4[\sqrt{H}] + 2} \leq \frac{1}{2} [H_*(M + N + 1)$$

$$+ \min\{(N + 1)(H - H_*), M(H - H_*), \frac{1}{3}(M(H - H_*) + 2M(N + 1))\}].$$

[Hint]: For $H \geq 2$, it is enough to show that $2H + \sqrt{H} \leq \frac{1}{3}(H + 4)(2\sqrt{H} + 1)$. Note that both sides can be divided by $2\sqrt{H} + 1$.

95. Derive Eq. (9.19) from Eq. (9.18). Also, referring to the Stan manual, explain the details of lines 40 and 53 in the Stan code.
96. What does the function 1q_decomp output when given a matrix A? Investigate the meanings of the functions qr, qr.Q, and qr.R and explain each step.

```
1   1q_decomp <- function(A) {
2     QR_decomp <- qr(t(A))
3     L <- t(qr.R(QR_decomp))
4     Q <- t(qr.Q(QR_decomp))
```

```
5     return(list(L = L, Q = Q))
6   }
7
8   A <- matrix(c(1, 2, 3, 4, 5, 6), nrow = 2, ncol = 3)
9   result <- lq_decomp(A)
10  L <- result$L
11  Q <- result$Q
```

[Hint]: There exist an orthogonal matrix Q and an upper triangular matrix R such that $A^\top = QR$, so if we set $L := R^\top$ (lower triangular matrix) and $S := Q^\top$ (orthogonal matrix), we can write $A = R^\top Q^\top = LS$.

97. From the values of $l(s)$ obtained in the appendix, derive Proposition 46. Also, even if the s of $\Phi^{(s)}$ in (9.24) is updated to the maximum value, a normal cross cannot be obtained. Why?

98. Referencing the Stan code in model14.stan, generalize model17.stan so that the K elements each have a common standard deviation $\sigma > 0$ instead of variance 1. Then, confirm that the same results are output when applying model14.stan as Stan code for the case of $K = 2$.

99. Show invariance of Jeffreys' prior for $\theta = t^4$ in Example 83.

100. In the proof of Proposition 47, when selecting the indices $(i, j) = (i_1, j_1), \ldots,$ (i_d, j_d) of $I_{i,j}(\theta)$ in such a way that there are no duplicates in $i_1, \ldots, i_d$ and $j_1, \ldots, j_d$, respectively, why can their products be multiplied like in Eq. (9.27)?

Bibliography

1. Zellner, A.: Bayesian and non-bayesian analysis of the regression model with multivariate student-t error terms. J. Am. Stat. Assoc. **71**(354), 400–405 (1976)
2. Leung, D., Drton, M.: Order-invariant prior specification in bayesian factor analysis. Stat. Probab. Lett. **111**, 60–66 (2016)
3. Files, B.T., Strelioff, M., Bonnevie, R.: Bayesian Reduced-Rank Regression with Stan. Technical report, CCDC Army Research Laboratory (2005). ARL-TR-8741
4. Heisuke, H.: Resolution of singularities of an algebraic variety over a field of characteristic zero I. Ann. Math. **79**(1), 109–203 (1964)
5. Heisuke, H.: Resolution of singularities of an algebraic variety over a field of characteristic zero II. Ann. Math. **79**(1), 205–326 (1964)
6. Miki, A., Sumio, W.: Resolution of singularities and generalization error with bayesian estimation for layereed neural network. IEICE Trans. **J88-DII**(10), 2112–2124 (2005)
7. Aoyagi, M., Watanabe, S.: Stochastic complexity of reduced rank regression in bayesian estimation. Neural Networks **18**, 924–933 (2005)
8. Matthew, D.H., Andrew, G.: The No-U-Turn sampler: adaptively setting path lengths in Hamiltonian Monte Carlo. J. Mach. Learn. Res. **15**(1), 1593–1623 (2014)
9. Watanabe, S.: Asymptotic equivalence of bayes cross validation and widely applicable information criterion in singular learning theory. J. Mach. Learn. Res. **11**, 3571–3594 (2010)
10. Watanabe, S.: A widely applicable bayesian information criterion. J. Mach. Learn. Res. **14**, 867–897 (2013)
11. Watanabe, S.: Learning efficiency of redundant neural networks in bayesian estimation. IEEE Trans. Neural Networks **12**(6), 1475–1486 (2001)
12. Sumio, W.: Algebraic Geometry and Statistical Learning Theory. Cambridge University Press (2009)
13. Sumio, W.: Theory and Methods of Bayesian Statiscics (in Japanese). Corona (2012)
14. Joe, S.: Statistical Learning with R: 100 Exercises for Building Logic. Springer (2020)
15. Shingo, M.: Tayo-tai, 2nd edn. Kyoritsu-Shuppan (1989)
16. Manabu, K.: Mathematical Statistics -Basis for Statistical Inference- (in Japanese). Kyoritsu-Shuppan Ltd (2020)

Index

Printed in the United States
by Baker & Taylor Publisher Services